A Practical Introduction to PSL

SERIES ON INTEGRATED CIRCUITS AND SYSTEMS

Anantha Chandrakasan, Editor
Massachusetts Institute of Technology
Cambridge, Massachusetts, USA

Published books in the series:

A Practical Guide for SystemVerilog Assertions
Srikanth Vijayaraghavan and Meyyappan Ramanathan
2005, ISBN 0-387-26049-8

Statistical Analysis and Optimization for VLSI: Timing and Power
Ashish Srivastava, Dennis Sylvester and David Blaauw
2005, ISBN 0-387-25738-1

Leakage in Nanometer CMOS Technologies
Siva G. Narendra and Anantha Chandrakasan
2005, ISBN 0-387-25737-3

Thermal and Power Management of Integrated Circuits
Arman Vassighi and Manoj Sachdev
2005, ISBN 0-398-25762-4

High Performance Energy Efficient Microprocessor Design
Vojin Oklobdzija and Ram Krishnamurthy (Eds.)
2006, ISBN 0-387-28594-6

Abstraction Refinement for Large Scale Model Checking
Chao Wang, Gary D. Hachtel and Fabio Somenzi
2006, ISBN 0-387-34155-2

Sub-threshold Voltage Circuit Design for Ultra-Low Power Systems
Alice Wang, Benton H. Calhoun, and Anantha Chandrakasan
2006, ISBN 0-387-33515-3

Cindy Eisner
Dana Fisman

A Practical Introduction to PSL

Cindy Eisner
IBM Haifa Research Laboratory
Haifa 31905 Israel

Dana Fisman
The Weizmann Institute of Science
Rehovot 76100 Israel

A Practical Introduction to PSL

Library of Congress Control Number: 2006928510

ISBN-10: 0-387-35313-5 e-ISBN-10: 0-387-36123-5
ISBN-13: 9780387353135 e-ISBN-13: 9780387361239

Printed on acid-free paper.

Printed in the United States of America.

9 8 7 6 5 4 3 2 1

springer.com

To the memory of my mother, Ruth Sorin Rosenbaum.
Your loving kindness and generosity of spirit
will stay with me forever.

C.E.

To Dr. Elsa Meryn Fisman, my mother and my guide.
Your endless love and devotion paved the way
for me to be the person I am today.

D.F.

Foreword

Functional verification is hard. Period. No disagreement here. But why is this so? Consider today's design flow: much of it is more or less automated, from RTL to netlist to layout to silicon. But all this automation depends upon having correct RTL input to start with, and there is little or no automation to help with RTL creation. It is hard enough for a designer to decide what RTL model he wants to build, and then to describe that RTL model correctly in a hardware description language. It is even more difficult for a verification engineer, who can't read the designer's mind, to verify that what the designer created not only represents the RTL model he had conceived, but also that the RTL model is an appropriate one for the problem at hand.

What makes RTL modeling and verification difficult is concurrency. It is easy to teach an engineer how to write procedural code that conforms to the synthesizable subset of a hardware description language. What is hard is understanding how the engineer's procedural code interacts with other components in the design over time. In fact, until recently we lacked effective languages to describe concurrent behaviors.

The IEEE 1850 Property Specification Language (PSL) is a language for the formal specification of concurrent systems. The language is particularly applicable for writing assertions about hardware designs. PSL supports multiple verification paradigms – including formal analysis, simulation, and acceleration/emulation. Furthermore, PSL assertions can be reused across multiple hierarchies of the design, ranging from block-level to system-level, providing verification consistency.

PSL is not just another assertion language; it is the industry's first and foremost standard property specification language – the result of over 5 years of intense collaboration among verification experts and practitioners. Based upon the Sugar language from IBM, PSL was initially developed in the Accellera Formal Verification Technical Committee (FVTC) and became an Accellera Standard in May 2003. The language was further refined within the IEEE 1850 Property Specification Language Working Group, resulting in the first IEEE standard assertion language, IEEE Std 1850™-2005 PSL, in September 2005. The only assertion language formally defined for multiple and mixed design language applications, PSL is widely used for assertion-based verification today.

Assertions are not just used for verification; they act as specifications as well. Today, assertions in design IP convey information about the usage requirements of that IP. Tomorrow, we can expect assertions to become more widely used in verification planning, as a means of specifying design features to test. And in the future, we can expect assertions to be used for defining the design itself. Just as gate-level design grew out of abstractions used to verify transistor-level netlists, and RTL design grew out of abstractions used to verify gate-level netlists, assertion-based design is likely to evolve out of assertion-based verification. So this book can help you understand not only one of today's leading-edge verification methodologies, but also the foundations for the design methodology of the future.

Cindy Eisner and Dana Fisman were the two key people who turned IBM Sugar into PSL. Their deep understanding of PSL's formal semantics was instrumental in both the Accellera and IEEE PSL standardization efforts. Cindy and Dana have now created the most authoritative source for information about PSL, designed to introduce the language incrementally in an easily understood fashion. *A Practical Introduction to PSL* provides a solid foundation for getting started with PSL today.

Harry Foster
Chair, Accellera Formal Verification Technical Committee (FVTC)
Chair, IEEE 1850 Property Specification Language (PSL) Working Group
Principal Engineer, Mentor Graphics Corporation

Erich Marschner
Co-Chair, Accellera Formal Verification Technical Committee (FVTC)
Secretary, IEEE 1850 Property Specification Language (PSL) Working Group
Senior Architect, Cadence Design Systems

Preface

This book describes the Property Specification Language PSL, recently standardized as IEEE Std 1850-2005. PSL provides a way to express properties of a design. For instance, we can state that every request receives an acknowledge, that every acknowledged request receives a grant within four to seven cycles unless the request is canceled first, that two consecutive writes should not be to the same address, or that when we see a read request with tag equal to `i`, then on the next four data transfers we expect to see a tag of `i`.

The primary audience is engineers involved in the design and verification of hardware, and we assume some familiarity with the basic vocabulary of hardware design, including such terms as *signal* and *clock cycle*, and with timing diagrams. Most of the examples are given in the Verilog flavor of PSL, but lack of familiarity with Verilog should not interfere with understanding the examples, which are only slightly affected by the flavor and are explained in detail.

PSL is a temporal logic, and a secondary audience is students of temporal logic. We expect that the intuitive descriptions of the temporal operators provided in the body of the book will make the formal semantics given in Appendix B accessible to beginning students, and that the illustrated examples will lead to a deeper understanding of the underlying issues.

Acknowledgments

We would like to thank Harry Foster, Erich Marschner and Yaron Wolfsthal for conceiving the idea of this book, and for their help and encouragement during the course of its writing.

Many people have contributed to our understanding of the underlying issues involved in defining a property specification language. For interesting and sometimes provocative discussions over the years, thank you to Roy Armoni, Ilan Beer, Shoham Ben-David, John Havlicek, Yoad Lustig, Anthony McIsaac,

Erich Marschner, Johan Mårtensson, Avigail Orni, Amir Pnueli, Ishai Rabinovitz, Yoav Rodeh, Sitvanit Ruah, David Van Campenhout, Moshe Vardi and Karen Yorav.

We thank the members of the Verification Guild for their excellent questions, many of which prompted text in this book.

For the generous contribution of their time reading and providing feedback on early drafts of this book, we would like to thank Andrea Fedeli, Oded Fuhrmann, Anthony McIsaac, Amir Nahir, Dejan Nickovic, Avigail Orni, Dmitry Pidan, Michal Rimon, Karen Yorav and Emmanuel Zarpas. Responsibility for any errors that might remain belongs of course to us. If you find such an error, please let us know. We would also welcome other comments and/or criticisms from readers.

Thanks also to Janick Bergeron, Roderick Bloem, Doron Bustan, Mike Gordon, Oded Kahana, Avner Landver, Orna Lichtenstein, Enrico Masala, Tama Mittleman, Nissim Ofek, Nir Piterman, Sivan Rabinovich, Chani Sacharen, Ohad Shacham, Ellen Yoffa and Avi Ziv.

For the enormous amount of work they collectively contributed to driving the standard that ultimately became PSL, we would like to thank Harry Foster, Erich Marschner, and the members of the Accellera Formal Verification Technical Committee and of the IEEE-1850 PSL working group.

Thank you to the IEEE for permission to reprint the BNF and the formal semantics of PSL from the IEEE Std 1850-2005.

We would like to acknowledge the support of the IBM Haifa Research Laboratory, where both of us are currently employed, and of the Weizmann Institute of Science, where Dana Fisman is a PhD student.

Thank you to our editor Carl Harris, to Deborah Doherty and the rest of the author support staff at Springer, and to Melissa Guasch for her excellent copyediting.

Finally, we would like to thank our families for their patience and support during the writing of this book.

Haifa, Israel
Rehovot, Israel
May 2006

Cindy Eisner
Dana Fisman

Contents

1

Introduction

PSL is a property specification language. It is a means to express *properties* of a design, and in addition to specify how verification tools should use those properties. For example, a property may be *asserted* – this specifies that the design in question is expected to behave as described by the property. A property may also be *assumed* – this specifies that the design in question expects its inputs to behave as described by the property. PSL also provides other directives, for instance a means to specify scenarios that should be *covered.*

PSL is much more than simply an assertion language. Nevertheless, assertions are at the heart of PSL, and many PSL users will use PSL only for its assertion capabilities. Thus, before we examine the complete language, a few words about what exactly a PSL assertion is are in order. Many programming languages (for example Java) and hardware description languages (for example VHDL) contain assert constructs. An assert construct provides the user with a way to check at run time or at simulation time that a certain condition holds, and to report a warning or an error if it does not. PSL assertions are similar in motivation, but much more extensive in their scope. In addition to simple Boolean conditions, a PSL assertion can contain *temporal* operators that allow the user to describe relations over time. For example, the PSL assertion `assert always (a -> next b);` specifies that whenever signal `a` holds, signal `b` must hold on the next cycle.

Java and VHDL assertions are designed to be embedded in executable code, and to be checked whenever execution reaches the point at which they appear. PSL assertions, on the other hand, typically stand alone, making a statement *about* the code without being a part of it. (While some tools support embedded PSL assertions, an embedded assertion is still not part of the code in the sense that there is no way to nest PSL assertions inside of executable statements. Embedded PSL assertions are located near the code they specify, but they are still about the code and not a part of it.)

PSL was designed to be mathematically rigorous, with the result that a PSL specification is both precise and automatically verifiable. Thus, a hardware specification written in PSL is machine readable and can be used as input to verification tools. In addition, PSL was designed to be easy to read and write, thus a PSL specification is human readable and can be used for documentation in order to clarify subtle points of an English specification.

The definitive definition of PSL can be found in IEEE Std 1850-2005. In this book, our goal is to give a more relaxed, user-friendly introduction to the language, as well as an in-depth discussion of its fine points. We do cover the whole of PSL, and for completeness have included as well the BNF and formal semantics as appendices.

The structure of PSL is based on four layers – the Boolean layer, the temporal layer, the verification layer and the modeling layer – and comes in a numbers of flavors, which influence the syntax in a limited way. The four layers of the language are:

- The *Boolean layer* is composed of Boolean expressions, that is, expressions whose value is either true or false. For example, `a` is a Boolean expression. PSL interprets a high signal as true, and a low signal as false, independent of whether the signal is active-high or active-low. Thus, if signal `a` is active-high, the Boolean expression `a` has the value true when `a` is asserted, and false when it is deasserted. And if signal `b` is active-low, the Boolean expression `b` has the value false when `b` is asserted and true when it is deasserted. In the remainder of this book, we will assume that all signals are active-high unless stated otherwise. Of course, it is easy to get the active-low versions of the example properties by switching `a` with `!a` and vice versa. The Boolean layer consists of any Boolean expression in the underlying flavor. For example, `a && !b` is a Boolean expression in the Verilog flavor stating that `a` holds and `b` does not, `a[31:0]==b[31:0]` is a Verilog-flavored Boolean expression stating that the 32-bit vectors `a[31:0]` and `b[31:0]` are equal, and `addr1[127:0]==128'b0` is a Verilog-flavored Boolean expression stating that the 128-bit vector `addr1[127:0]` has the value zero.
- The *temporal layer* consists of temporal properties which describe the relationships between Boolean expressions over time. For example, `always (req -> next ack)` is a temporal property expressing the fact that whenever (`always`) signal `req` is asserted, then (`->`) at the next cycle (`next`), signal `ack` is asserted.
- The *verification layer* consists of directives which describe how the temporal properties should be used by verification tools. For example, `assert always (req -> next ack);` is a verification directive that tells the tools to verify that the property `always (req -> next ack)` holds. Other verification directives include an instruction to assume, rather than verify, that a particular temporal property holds, or to specify coverage criteria. The verification layer also provides a means to group PSL statements into *verification units*.

```
vunit example1 {

   assert never (a and b);

}
```

(i) VHDL flavor

```
vunit example1 {

   assert never (a && b);

}
```

(ii) Verilog flavor

Fig. 1.1: The same vunit in two different flavors

- The *modeling layer* provides a means to model behavior of design inputs, and to declare and give behavior to auxiliary signals and variables. This part of the modeling layer is simply a subset of the underlying flavor. For instance, the declaration of auxiliary signals in the Verilog flavor follows Verilog syntax.

The five flavors of PSL correspond to the hardware description languages Verilog and VHDL, to the language GDL, the environment description language of the RuleBase model checker, and to SystemVerilog and SystemC. While the flavor has some influence over the syntax – for instance, it affects the syntax of Boolean expressions – the vast majority of the language is the same across flavors.

Figure 1.1 shows a PSL specification in the VHDL and Verilog flavors. In each case, the specification states that signals `a` and `b` are mutually exclusive. While a PSL user typically does not spend a lot of time thinking about the boundaries between the various layers, we will point them out in this first example. The Boolean expressions `a and b` in the VHDL flavor and `a && b` in the Verilog flavor belong to the Boolean layer and describe the situation in which both signal `a` and signal `b` are asserted. The keyword `never` belongs to the temporal layer and indicates that the Boolean expression is expected to hold in no cycle. Together, the temporal keyword `never` and the Boolean expressions `a and b` in the VHDL flavor or `a && b` in the Verilog flavor form a *property*. The keyword `assert` belongs to the verification layer and instructs the verification tool to check that the property holds in the design under test. The keyword `vunit` also belongs to the verification layer and introduces the name of the verification unit (`example1`). The modeling layer is not used in this example.

In the remainder of this book we use the Verilog flavor unless stated otherwise. We focus almost exclusively on the temporal layer, which is the heart of the language and makes extensive use of the Boolean layer. Chapter 10 briefly discusses various aspects of the Boolean, modeling and verification layers not covered elsewhere. Throughout, we will assume that `` `true `` has been defined to be `1'b1` (i.e., a Verilog expression that holds at every cycle) and that `` `false `` has been defined to be `1'b0` (i.e., a Verilog expression that does not hold at any cycle).

2

Basic Temporal Properties

While the Boolean layer consists of Boolean expressions that hold or do not hold at a given cycle, the temporal layer provides a way to describe relationships between Boolean expressions over time. A PSL assertion typically looks in only one direction – forwards from the first cycle (although it is possible to look backwards using built-in functions such as `prev()`, `rose()` and `fell()`). Thus, the simple PSL assertion `assert a;` states that `a` should hold at the very first cycle, while the PSL assertion `assert always a;` states that `a` should hold at the first cycle and at every cycle following the first cycle – that is, at every cycle.

By combining the temporal operators in various ways we can state properties such as "every request receives an acknowledge", "every acknowledged request receives a grant within four to seven cycles unless the request is canceled first", "two consecutive writes should not be to the same address", and "when we see a read request with tag equal to `i`, then on the next four data transfers we expect to see a tag of `i`".

The temporal layer is composed of the Foundation Language (FL) and the Optional Branching Extension (OBE). The FL is used to express properties of single traces, and can be used in either simulation or formal verification. The OBE is used to express properties referring to sets of traces, for example "there exists a trace such that ...", and is used in formal verification. In this book we concentrate on the Foundation Language.

The Foundation Language itself is composed of two complementary styles – LTL style, named after the temporal logic LTL on which PSL is based, and SERE style, named after PSL's Sequential Extended Regular Expressions, or SEREs. In this chapter we present the basic temporal operators of LTL style. We provide only a taste – enough to get the basic idea and to give some context to the philosophical issues that we discuss next.

Throughout this book, we make extensive use of examples. Each example property or assertion and its associated timing diagram (which we term a *trace*) are grouped together in a figure. Such a figure will contain one or more traces numbered with a parenthesized lower case Roman numeral, and one or more properties numbered by appending a lower case letter to the figure number. For instance, Figure 2.1 contains Trace 2.1(i) and Assertions 2.1a, 2.1b, and 2.1c.

2.1 The `always` and `never` operators

We have already seen the basic temporal operators `always` and `never`. Most PSL properties will start with one or the other. This is because a "bare" (Boolean) PSL property refers only to the first cycle of a trace. For example, Assertion 2.1a requires only that the Boolean expression `!(a && b)` hold at the first cycle. Thus, Assertion 2.1a holds on Trace 2.1(i) because the Boolean expression `!(a && b)` holds at cycle 0. In order to state that we want it to hold at every cycle of the design, we must add the temporal operator `always` to get Assertion 2.1b. Assertion 2.1b does not hold on Trace 2.1(i) because the Boolean expression `!(a && b)` does not hold at cycle 5. Equivalently, we could have swapped the `always` operator and the Boolean negation `!` with `never`, to get Assertion 2.1c.

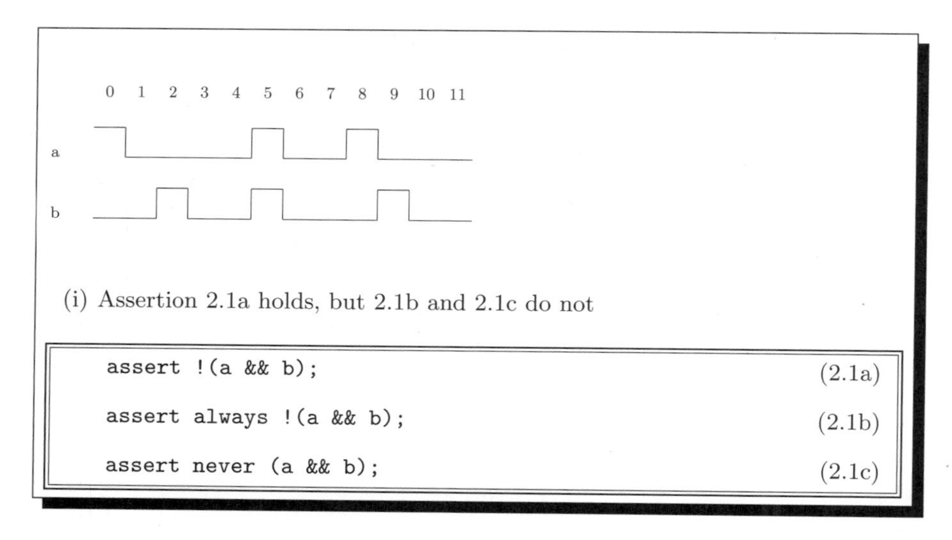

(i) Assertion 2.1a holds, but 2.1b and 2.1c do not

```
assert !(a && b);                (2.1a)

assert always !(a && b);         (2.1b)

assert never (a && b);           (2.1c)
```

Fig. 2.1: The `always` and `never` operators

Both Assertion 2.1b and Assertion 2.1c state that signals `a` and `b` are mutually exclusive. Obviously, anything that can be stated with the `always` operator can be stated with the `never` operator and vice versa, simply by negating the operand when switching between `always` and `never`. PSL provides both operators for convenience, as sometimes it is more natural to state the property in the positive (that is, stating what must hold at all cycles) and sometimes in the negative (that is, what must not hold for any cycle). In general, there are many ways to state any property in PSL. We will see other examples throughout the rest of this book.

2.2 The `next` operator

Another temporal operator is the `next` operator. It indicates that the property will hold if its operand holds at the next cycle. For instance, Assertion 2.2a states that whenever `a` holds, then `b` should hold in the next cycle. Assertion 2.2a uses another important operator, the logical implication operator (`->`). While the logical implication operator is a Boolean and not a temporal operator (it does not link two sub-properties over time), it plays a very important role in many temporal properties. A logical implication `prop1 -> prop2` holds if either `prop1` does not hold or `prop2` holds. A good way to think of it is like an if-then expression, with the else-part being implicitly true. That is, `prop1 -> prop2` can be understood as "if prop1 then prop2 else true". Thus, the sub-property `a -> next b` in our example holds if either `a` does not hold (because then the property defaults to true) or if `a` holds and also `next b` holds. By adding an `always` operator, we get a property that holds if the sub-property `a -> next b` holds at every cycle. Thus, Assertion 2.2a states that whenever `a` holds, `b` must hold at the next cycle. Assertion 2.2a holds on Trace 2.2(i) because every assertion of signal `a` is followed by an assertion of signal `b`. This is shown in the "if" and "then" annotations on Trace 2.2(ii). The "additional" assertions of signal `b` at cycles 1 and 10 are allowed by Assertion 2.2a, because it says nothing about the value of `b` in cycles other than those following an assertion of `a`.

Note that the cycles involved in satisfying one assertion of signal `a` may overlap with those involved in satisfying another assertion. For example, consider Trace 2.2(iii), which is simply Trace 2.2(ii) with the if-then pairs numbered. There are four assertions of signal `a` on Trace 2.2(iii), and thus four associated cycles in which `b` must be asserted. Each pair of cycles (an assertion of `a` followed by an assertion of `b`) is numbered in Trace 2.2(iii). Consider pairs 2 and 3. Signal `a` is asserted at cycle 4 in pair 2, thus signal `b` needs to be asserted at cycle 5 in order for Assertion 2.2a to hold. Signal `a` is asserted at cycle 5 in pair 3, thus requiring that signal `b` be asserted at cycle 6. Pairs 2 and 3 overlap, because while we are looking for an assertion of signal `b` at cycle 5 in order to satisfy the assertion of `a` at cycle 4, we see an additional assertion of signal `a` that must be considered.

Assertion 2.2a does not hold on Trace 2.2(iv) because the third assertion of signal `a`, at cycle 5, is missing an assertion of signal `b` at the following cycle.

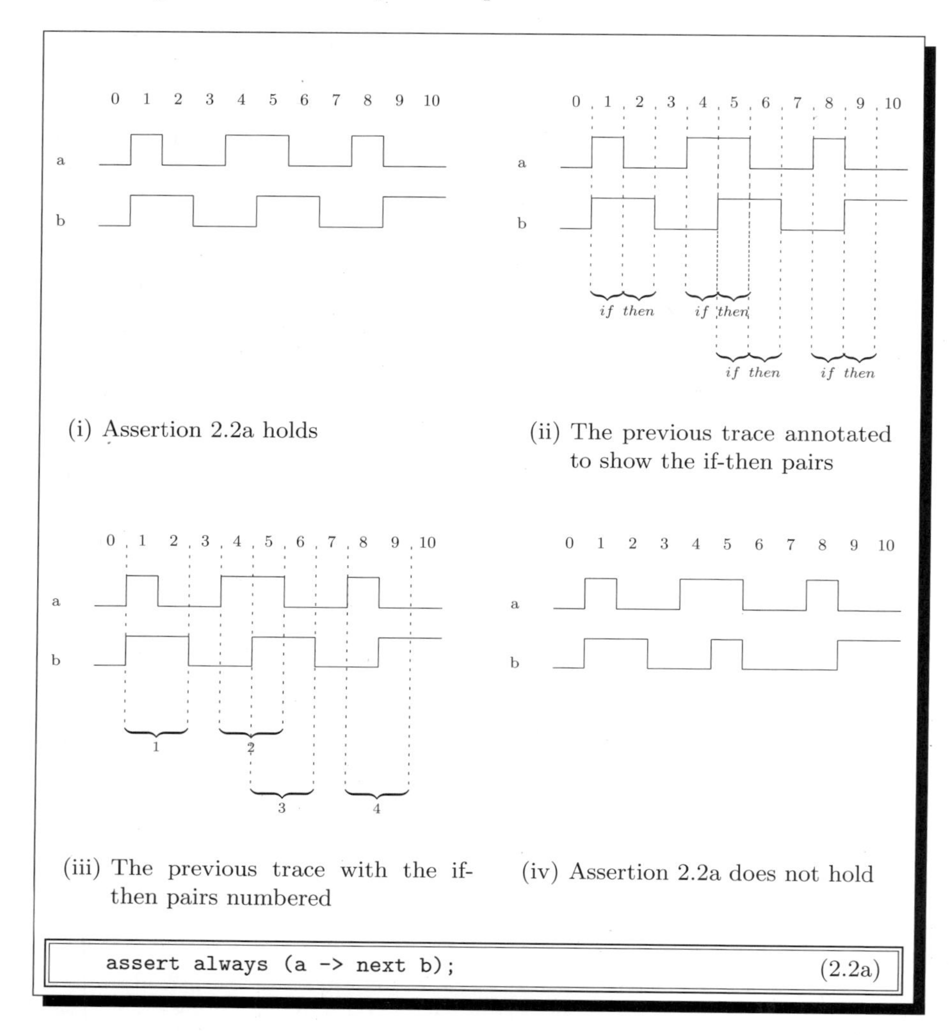

Fig. 2.2: The `next` and logical implication operators

2.3 Variations on `next` including `next_event`

A `next` property holds if its operand holds in the next cycle. Variations on the `next` operator allow you to specify the nth next cycle, and ranges of future cycles. A `next[n]` property holds if its operand holds in the nth next cycle. For example, Assertion 2.3a states that whenever signal `a` holds, signal `b` holds three cycles later. Assertion 2.3a holds on Traces 2.3(i), 2.3(iii), and 2.3(iv), while it does not hold on Traces 2.3(ii) or 2.3(v) because of a missing assertion of signal `b` at cycle 7, and does not hold on Trace 2.3(vi) because of a missing assertion of signal `b` at cycle 5.

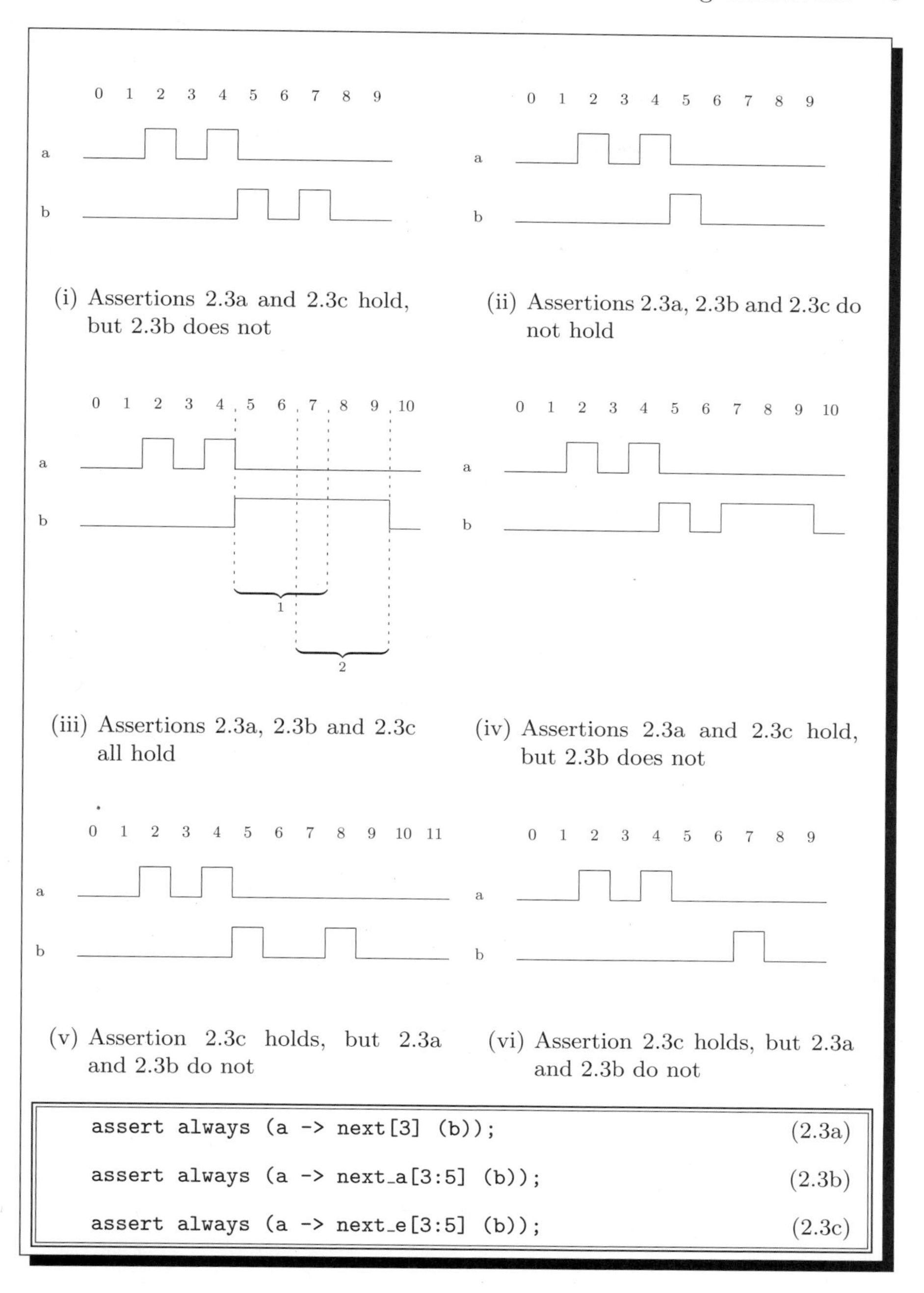

Fig. 2.3: The next[n], next_a[i:j] and next_e[i:j] operators

A `next_a[i:j]` property holds if its operand holds in *all* of the cycles from the ith next cycle through the jth next cycle, inclusive. For example, Assertion 2.3b states that whenever signal `a` holds, signal `b` holds three, four and five cycles later. It holds on Trace 2.3(iii) and does not hold on Traces 2.3(i), 2.3(ii), 2.3(iv), 2.3(v), or 2.3(vi).

Previously we discussed the fact that the cycles involved in satisfying one assertion of signal `a` may overlap those involved in satisfying another assertion of `a`. Trace 2.3(iii) has been annotated to emphasize this point for Assertion 2.3b. Signal `b` must be asserted in cycles 5 through 7 (marked as "1") because of the assertion of `a` at cycle 2, and must be asserted in cycles 7 through 9 (marked as "2") because of the assertion of `a` at cycle 4.

A `next_e[i:j]` property holds if there *exists* a cycle from the next `i` through the next `j` cycles in which its operand holds. For example, Assertion 2.3c states that whenever signal `a` holds, signal `b` holds either three, four, or five cycles later. There is nothing in Assertion 2.3c that prevents a single assertion of signal `b` from satisfying multiple assertions of signal `a`, thus it holds on Trace 2.3(vi) because the assertion of `b` at cycle 7 comes five cycles after the assertion of signal `a` at cycle 2, and three cycles after the assertion of signal `a` at cycle 4. We examine the issue of specifying a one-to-one correspondence between signals in Section 13.4.2.

Assertion 2.3c also holds on Traces 2.3(i), 2.3(iii), 2.3(iv), and 2.3(v), since there are enough assertions of signal `b` at the appropriate times. In Traces 2.3(i), 2.3(iii), and 2.3(iv) there are more than enough assertions of `b` to satisfy the property being asserted (in Trace 2.3(i), the assertion of `b` at cycle 7 is enough, because it comes five cycles after the assertion of `a` at cycle 2, and three cycles after the assertion of `a` at cycle 4). In Trace 2.3(v) there are just enough assertions of `b` to satisfy the requirements of Assertion 2.3c.

The `next_event` operator is a conceptual extension of the `next` operator. While `next` refers to the next cycle, `next_event` refers to the next cycle in which some Boolean condition holds. For example, Assertion 2.4a expresses the requirement that whenever a high priority request is received (signal `high_pri_req` is asserted), then the next grant (assertion of signal `gnt`) must be to a high priority requester (signal `high_pri_ack` is asserted). Assertion 2.4a holds on Trace 2.4(i). There are two assertions of signal `high_pri_req`, the first at cycle 1 and the second at cycle 10. The associated assertions of `gnt` occur at cycles 4 and 11, respectively, and `high_pri_ack` holds in these cycles.

The `next_event` operator includes the current cycle. That is, an assertion of `b` in the current cycle will be considered the next assertion of `b` in the property `next_event(b)(p)`. For instance, consider Trace 2.4(ii). Trace 2.4(ii) is similar to Trace 2.4(i) except that there is an additional assertion of `high_pri_req` at cycle 8 and two additional assertions of `gnt` at cycles 8 and 9, one of which has an associated `high_pri_ack`. Assertion 2.4a holds on Trace 2.4(ii) because the assertion of `gnt` at cycle 8 is considered the next assertion of `gnt` for the assertion of `high_pri_req` at cycle 8. If you want to

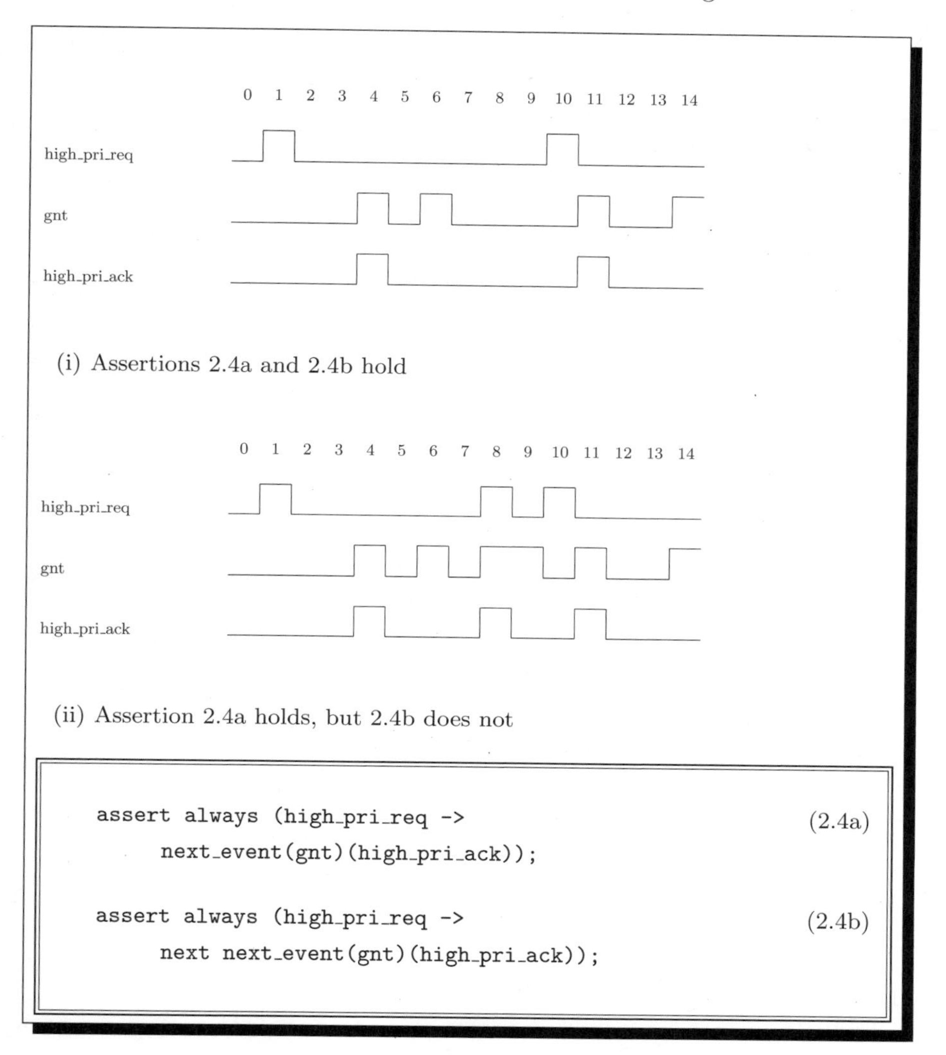

Fig. 2.4: `next_event`

exclude the current cycle, simply insert a `next` operator in order to move the current cycle of the `next_event` operator over by one, as in Assertion 2.4b. Assertion 2.4b does not hold on Trace 2.4(ii). Because of the insertion of the `next` operator, the relevant assertions of `gnt` have changed from cycles 4, 8 and 11 for Assertion 2.4a to cycles 4, 9 and 11 for Assertion 2.4b, and at cycle 9 there is no assertion of `high_pri_ack` in Trace 2.4(ii).

Just as we can use `next[i]` to indicate the ith next cycle, we can use `next_event(b)[i]` to indicate the ith occurrence of `b`. For example, in order

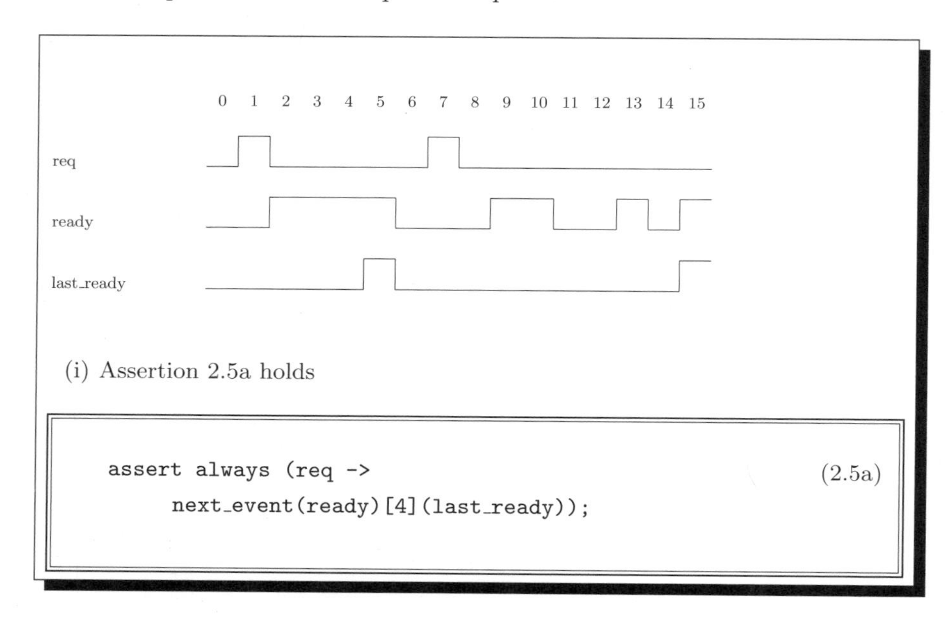

Fig. 2.5: `next_event[n]`

to express the requirement that every time a request is issued (signal `req` is asserted), signal `last_ready` must be asserted on the fourth assertion of signal `ready`, we can code Assertion 2.5a. Assertion 2.5a holds on Trace 2.5(i). For the first assertion of `req`, at cycle 1, the four assertions of `ready` happen to come immediately and in consecutive cycles. For the second assertion of `req`, at cycle 7, the four assertions of `ready` do not happen immediately and do not happen consecutively either – they are spread out over seven cycles, interspersed with cycles in which `ready` is deasserted. However, the point is that in both cases, signal `last_ready` is asserted on the fourth assertion of `ready`, thus Assertion 2.5a holds on Trace 2.5(i).

As with `next_a[i:j]` and `next_e[i:j]`, the `next_event` operator also comes in forms that allow it to indicate *all* of a range of future cycles, or the *existence* of a future cycle in such a range. The form `next_event_a(b)[i:j](f)` indicates that we expect `f` to hold on all of the ith through j^{th} occurrences of `b`. For example, Assertion 2.6a indicates that when we see a read request (assertion of signal `read_request`) with tag equal to `i`, then on the next four data transfers (assertion of signal `data`), we expect to see tag `i`. Assertion 2.6a uses the `forall` construct, which we will examine in detail later. For now, suffice it to say that Assertion 2.6a states a requirement that must hold for all possible values of the signal `tag[2:0]`. Assertion 2.6a holds on Trace 2.6(i) because after the first assertion of signal `read_request`, where `tag[2:0]` has the value 4, the value of `tag[2:0]` is also 4 on the next four assertions of signal `data` (at cycles 5, 9, 10 and 11). Likewise, on the second assertion of

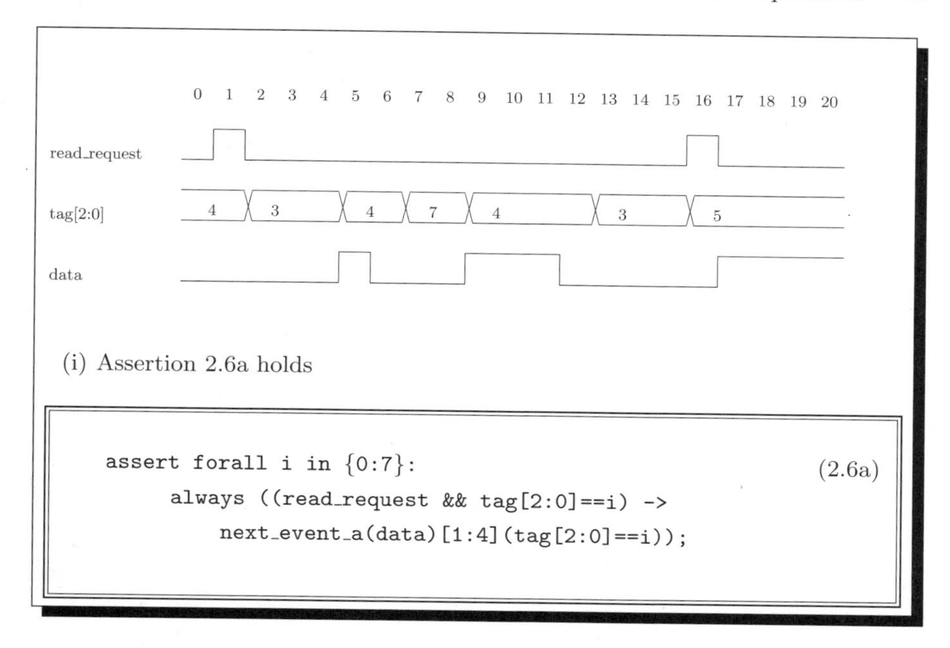


```
assert forall i in {0:7}:                                    (2.6a)
    always ((read_request && tag[2:0]==i) ->
        next_event_a(data)[1:4](tag[2:0]==i));
```

Fig. 2.6: next_event_a[i:j]

signal read_request, where tag[2:0] has the value 5, the value of tag[2:0] is also 5 on the next four assertions of signal data (at cycles 17 through 20).

In order to indicate that we expect something to happen on one of the next i^{th} to j^{th} cycles, we can use the next_event_e(b)[i:j](f) operator, which indicates that we expect f to hold on one of the ith through j^{th} occurrences of b. For example, consider again Assertion 2.4a. It requires that whenever a high priority request is received, the next grant must be to a high priority requester. Suppose instead that we require that one of the next *two* grants be to a high priority requester. We can express this using Assertion 2.7a. Assertion 2.7a holds on Trace 2.7(i) because every time that signal high_pri_req is asserted, signal high_pri_ack is asserted on one of the next two assertions of gnt.

The syntax of the range specification for all operators – including those we have not yet seen – is flavor dependent. In the Verilog, SystemVerilog and SystemC flavors, it is [i:j]. In the VHDL flavor it is [i to j]. In the GDL flavor it is [i..j].

2.4 The until and before operators

The until operator provides another way to move forward, this time while putting a requirement on the cycles in which we are moving. For example, Assertion 2.8a states that whenever signal req is asserted, then, starting at

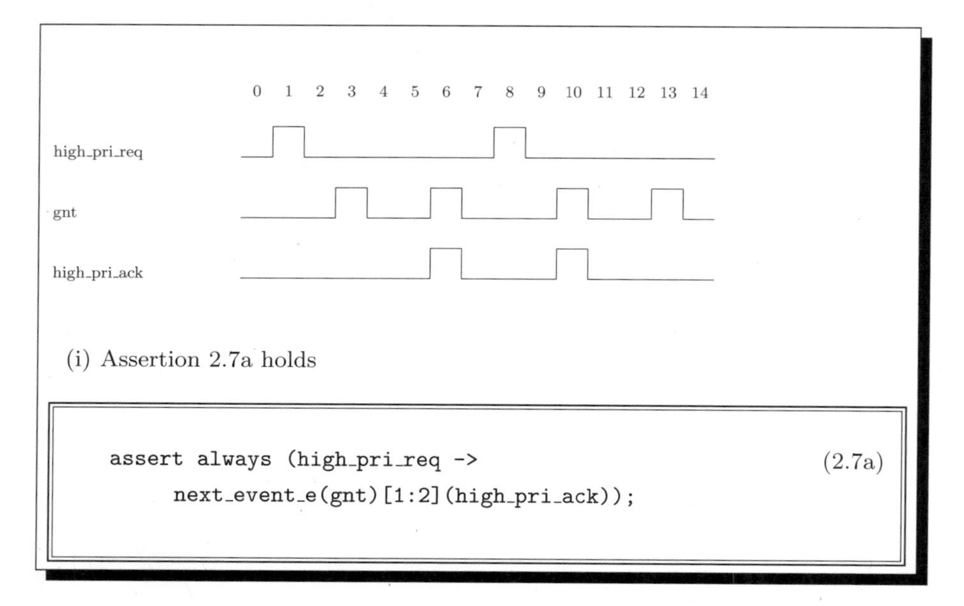

Fig. 2.7: `next_event_e[i:j]`

the next cycle, signal `busy` must be asserted up until signal `done` is asserted. Assertion 2.8a requires that `busy` will be asserted up to, but not necessarily including, the cycle where `done` is asserted. In order to include the cycle where `done` is asserted, use the operator `until_`. The underscore (_) is intended to represent the extra cycle in which we require that `busy` should stay asserted, so Assertion 2.8b states that whenever signal `req` is asserted, then starting from the next cycle, `busy` must be asserted and must stay asserted up until and including the cycle where `done` is asserted. For example, Assertion 2.8a holds on Trace 2.8(i), but Assertion 2.8b does not, because `busy` is not asserted at cycles 4 and 10. Both Assertions 2.8a and 2.8b hold on Trace 2.8(ii): Assertion 2.8a does not prohibit the assertion of `busy` at cycles 4 and 10 – it just does not require it.

If signal `done` is asserted the cycle after signal `req` is asserted, Assertion 2.8a does not require that signal `busy` be asserted at all, while Assertion 2.8b does. That is, Assertion 2.8a holds on Trace 2.8(iii) – the fact that `done` happens immediately after `req` leaves no cycles on which busy needs to be asserted. Assertion 2.8b does not hold on Trace 2.8(iii) because of a missing assertion of `busy` in the cycle in which `done` is asserted.

The `before` family of operators provides an easy way to state that we require some signal to be asserted before some other signal. For example, suppose that we have a pulsed signal called `req`, and we have the requirement that before we can make a second request, the first must be acknowledged. We can express this in PSL using Assertion 2.9a. We need the `next` to take us forward

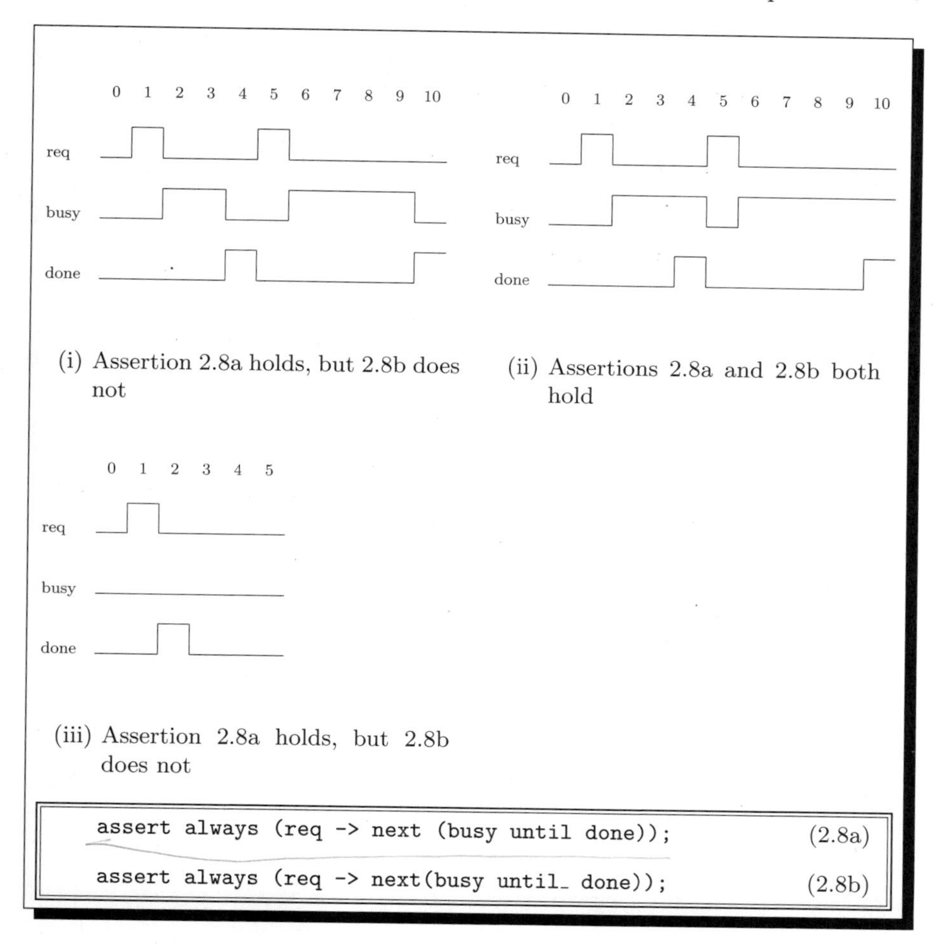

Fig. 2.8: The `until` and `until_` operators

one cycle so that the `req` in `(ack before req)` is sure to refer to some other `req`, and not the one we have just seen. To understand this, let us examine a flawed version of the same specification. Consider Assertion 2.9b. It requires that `(ack before req)` hold at every cycle in which `req` holds. Consider, for example, cycle 1 of Trace 2.9(i). Signal `req` is asserted. Therefore, `(ack before req)` must hold at cycle 1. However, it does not, because starting at cycle 1 and looking forward, we first see an assertion of signal `req` (at cycle 1), and only afterwards an assertion of signal `ack` (at cycle 3) – so `req` is asserted before `ack`, and not the other way around. Assertion 2.9a, on the other hand, states what we want: at cycle 1, for example, we require `next (ack before req)` to hold. Therefore, we require that `(ack before req)` hold at cycle 2.

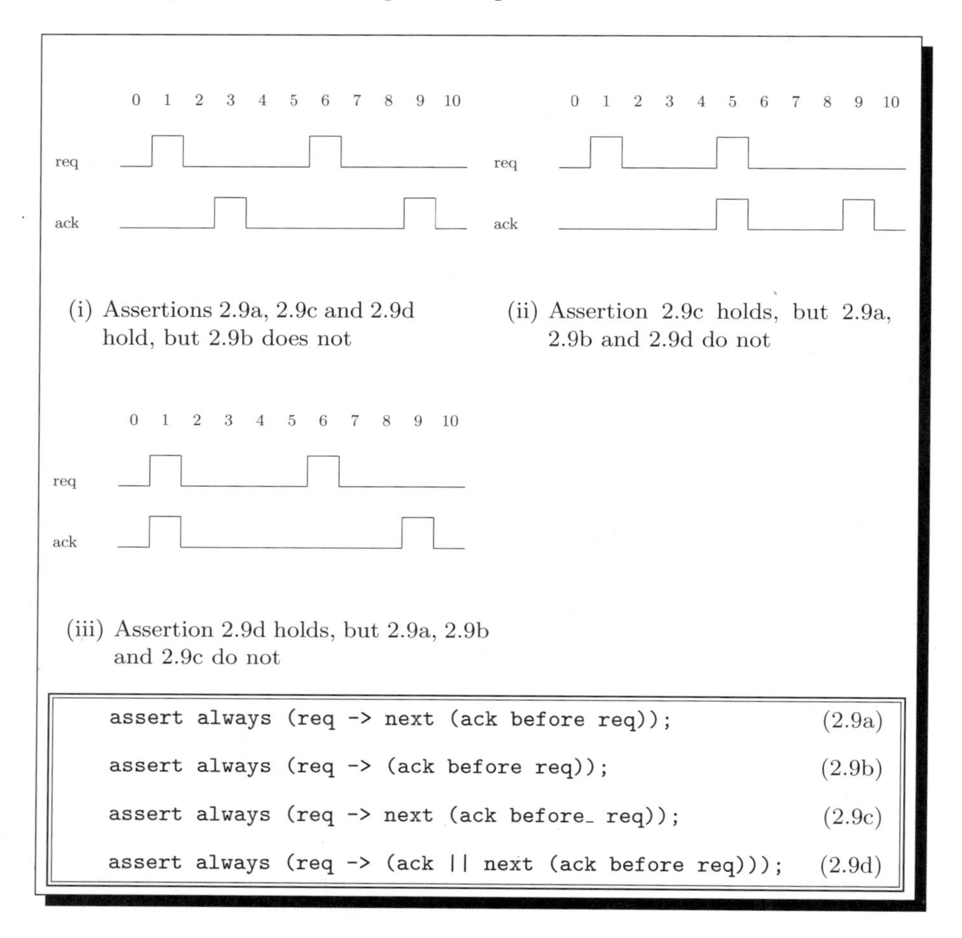

Fig. 2.9: The `before` and `before_` operators

Starting at cycle 2 and looking forward, we first see an assertion of `ack` (at cycle 3), and only afterwards an assertion of `req` (at cycle 6).

The `before` operator requires that its first operand happen strictly before its second. In order to specify that something must happen before or at the same cycle as something else, use `before_`. The underscore (_) is intended to represent the cycle in which we allow an overlap between the left and right sides. For example, in order to specify that behavior like that shown in Trace 2.9(i) is allowed, and that in addition behavior like that shown in Trace 2.9(ii) is allowed, use Assertion 2.9c.

What if the assertion of `ack` is allowed to come, not together with the next assertion of `req`, but rather together with the request being acknowledged? In other words, what if in addition to the behavior shown in Trace 2.9(i), we want to allow the behavior shown in Trace 2.9(iii)? As we have seen previously, Assertion 2.9b is not the answer. Rather, we can code Assertion 2.9d.

2.5 eventually!

The eventually! operator allows you to specify that something must occur in the future without saying exactly when. For example, Assertion 2.10a states that every request (assertion of req) must be followed at some time with an acknowledge (assertion of ack). There is nothing in Assertion 2.10a to prevent a single acknowledge from satisfying the requirement for multiple requests, thus Assertion 2.10a holds on Trace 2.10(i). We examine the issue of specifying a one-to-one correspondence between signals in Section 13.4.2.

The exclamation point (!) of the eventually! operator indicates that it is a strong operator. We discuss weak vs. strong temporal operators in detail in Chapter 4.

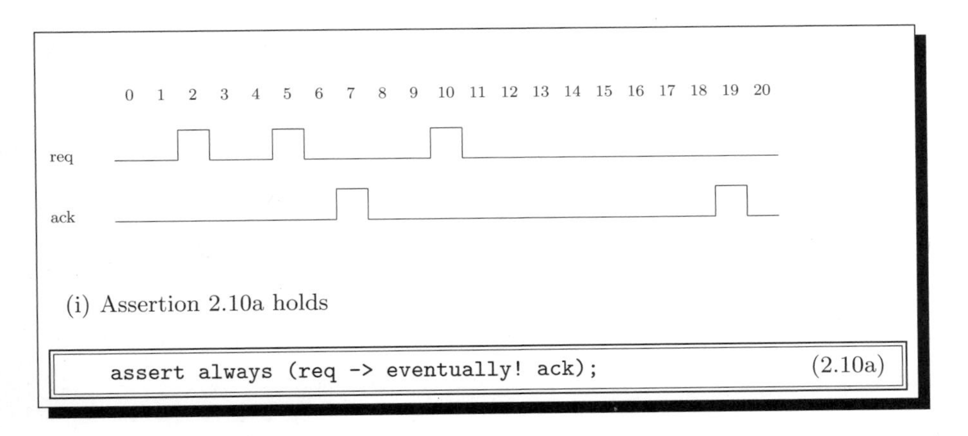

Fig. 2.10: The eventually! operator

3

Some Philosophy

We have seen some basic PSL and gotten a feel for how it is intended to be used. Before we continue, we discuss some of the concepts at the root of PSL.

3.1 Assertions vs. properties

As we have seen, a PSL assertion is made up of the keyword `assert` plus the PSL *property* being asserted, followed by a semi-colon. For example, in the assertion `assert always (a -> next b);`, the property is `always (a -> next b)`. A property holds or does not hold on a given trace. It is agnostic about whether or not holding is a good thing. An assertion, on the other hand, tells the verification tool that the property being asserted is required to hold.

In the remainder of this book we will be very careful about distinguishing between a property, which merely describes behavior, and an assertion, which sets a requirement. For example, we might say that an assertion `assert always (a -> next b);` requires that whenever signal `a` is asserted, signal `b` will be asserted in the next cycle. However, we will never say that about the property `always (a -> next b)`, for two reasons. First, because if the property is used in an assumption (`assume always (a -> next b);`), then no requirement is being stated. Second, because a property may be used as a sub-property of a more complicated property, and as such, it does not state a requirement on its own.

3.2 The notion of time

When a Boolean assertion is embedded in code that is being run, like in the simple assertions of Java and (simulated) VHDL, the notion of time need not be defined – an assertion is checked whenever the statement containing the assertion is executed. For the more complicated assertions of PSL, which first of all stand apart from the code (so that the notion of "execution" is foreign to them) and which second of all span multiple time steps, the notion of time must be given more consideration.

PSL assumes that time is discrete, that is, that time consists of a sequence of evaluation cycles. The meaning of a PSL property is defined relative to such a sequence of cycles. In this book, we will refer to such a sequence of cycles as a *trace.*

PSL does not dictate how time ticks – that is, it does not dictate how such a sequence of cycles is extracted from a design under verification. This means that the sequence of cycles as seen by two verification tools is not necessarily the same. For example, a cycle-based simulator sees a sequence of signal values calculated cycle-by-cycle, while an event-based simulator running on the same design sees a more detailed sequence of signal values.

Nevertheless, there is a way to ensure that the meaning of a PSL property is not affected by the granularity of time as seen by the verification tool. PSL properties can be modified by using a *clock expression* to indicate that time should be measured in clock cycles of the clock expression. In the case of a clocked property, the result of a cycle-based tool will be the same as the result of an event-based tool. PSL allows the specification of a *default clock*, so that the clock does not have to be mentioned explicitly for each and every property. In most of this book we have assumed a singly clocked design under the cycle-based model, and thus most examples omit the explicit mention of the clock. Clocks are discussed in detail in Chapters 6 and 14.

3.3 Designs and traces

The purpose of a PSL property is to describe the desired behavior of a design. In order to do so, it is usually convenient to examine individual traces of that design. The Foundation Language (FL), which we focus on in this book, uses this approach, and thus throughout most of this book we shall be interested in whether or not a particular PSL property holds on a particular trace. The Foundation Language is suitable for both static (formal) and dynamic (simulation-based) verification.

Another approach, used by the Optional Branching Extension (OBE), uses a tree structure that represents multiple paths. This approach is applicable only to formal verification, and is touched on very briefly in Chapter 11.

3.4 Current cycle, sub-traces, and modularity

When a PSL property composed of two or more sub-properties is checked on a trace, it is sometimes necessary to decide the meaning of the sub-properties on a sub-trace. The *current cycle* is the name we give to the first cycle of a trace or a sub-trace on which we are evaluating a property or a sub-property.

Assuming that the cycles of a trace are numbered starting from 0, the current cycle of an assertion is 0. The current cycle of a sub-property of some enclosing property depends on its use in the enclosing property. The operands of a Boolean operator have the same current cycle as the parent property. The operands of a temporal operator have a current cycle that is related to the current cycle of the parent property in a way dictated by the temporal operator. For example, the `next` operator increases the current cycle by one, the `always` operator creates "multiple instances" of the sub-property, each of which has a current cycle that corresponds either to the current cycle or to some future cycle, etc.

NOTE: It is very important to understand that the term "multiple instances" is intended to convey an intuition which is useful in understanding the `always` operator. It is *not* intended to hint that any actual instantiation is taking place; neither does it imply that a tool implementing PSL needs to create multiple instances of an assertion checker, spawn multiple instances of a process, or in any other way cause the words "multiple instances" to correspond to actual instances of anything whatsoever. On the contrary, there are many efficient implementations of PSL in which the `always` operator does not create, spawn, or in any other way generate actual multiple instances. Still, the term "multiple instances" is a good way to gain intuition about how the `always` operator works, and a naive and inefficient implementation may well generate multiple instances of a checker or of a process.

Getting back to our main point, consider the assertion `assert always (a -> next b);`. The current cycle of `always (a -> next b)` is 0. Whether or not `always (a -> next b)` holds at cycle 0 depends on whether or not the sub-property `(a -> next b)` holds at every cycle from 0 onwards. The current cycle for a particular evaluation of the sub-property `(a -> next b)` will be some cycle N. Finally, in order to determine whether or not sub-property `(a -> next b)` holds at cycle N, we will need to evaluate sub-properties `a` and `next b` with a current cycle of N, which means that we need to evaluate sub-property `b` with a current cycle of $N + 1$.

To make the discussion more concrete, let's consider our assertion, Assertion 3.1a, on Trace 3.1(i). Signal `a` holds in cycles 4 and 8. Signal `b` holds in cycle 5, and therefore `next b` holds in cycle 4. This is shown in Trace 3.1(ii), an annotated version of Trace 3.1(i). Since `a` holds in cycles 4 and 8, and `next b` holds in cycle 4, we get that `(a -> next b)` holds in all cycles but cycle 8, as shown in Trace 3.1(ii). (Remember that the else-part defaults to true, so `(a -> next b)` holds in all cycles where `a` does not hold, and in addition in all cycles where `a` holds and `next b` does too.) The entire property `always (a -> next b)` therefore holds in cycles 9, 10, 11, 12 and 13 (because in these cycles, `(a -> next b)` holds "now" and in all future cycles). Thus, Assertion 3.1a does not hold on Trace 3.1(i) (because it does not hold on cycle 0 – the first cycle of the trace).

The above explanation seems straightforward. However, the idea of modularity hides some subtle points – for example, that the value of the prop-

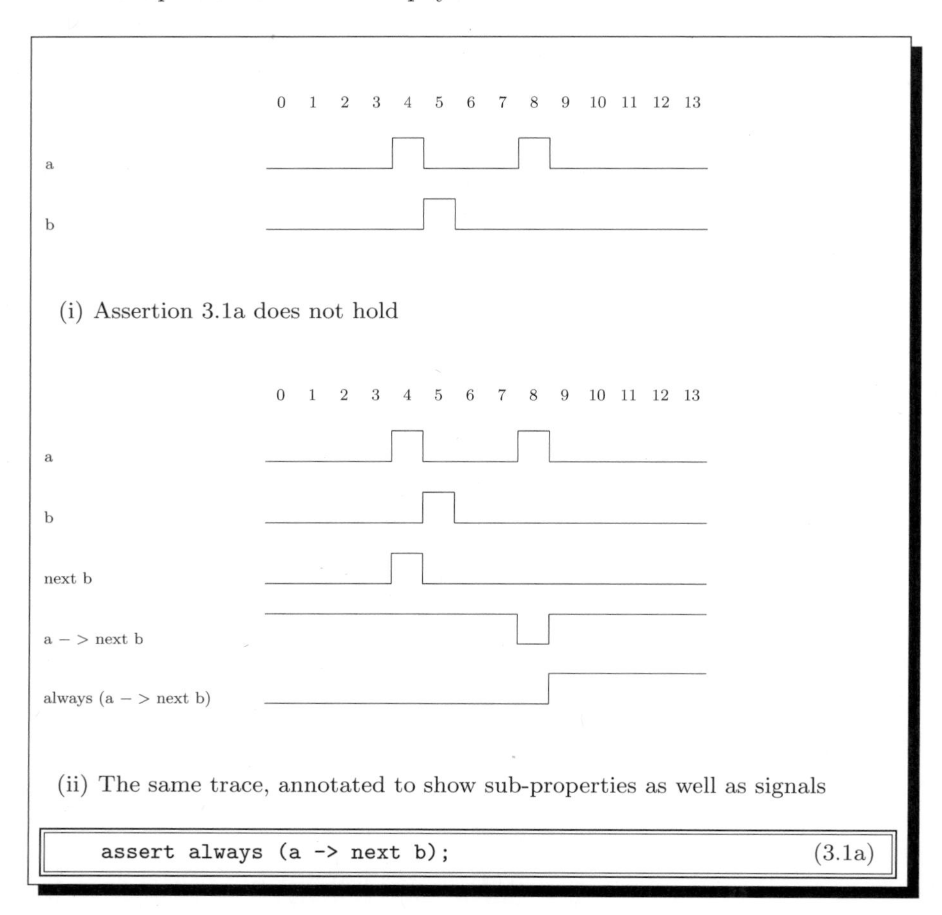

Fig. 3.1: A concrete example

erty `a` is dependent only on the current cycle. Thus, Assertion 3.2a holds on Trace 3.2(i), because signal `a` holds in cycle 0. If you want to express the fact that `a` should hold in every cycle, you must use the `always` operator, as in Assertion 3.2b. Assertion 3.2b does not hold on Trace 3.2(i).

Another subtle point is that a property can hold on a suffix of a trace without holding on the trace itself. Thus, the property `always a` holds on the sub-traces of Trace 3.2(i) starting at cycles 12, 13, and 14. This would be important if we used the property `always a` as a sub-property of some other property. For example, Assertion 3.2c holds on Trace 3.2(i) because the property `always a` holds on the 12th next cycle.

Finally, the cycles involved in calculating the left-hand side of a logical implication may overlap those involved in calculating the right-hand side of a logical implication. For example, consider Assertion 3.3a on Trace 3.3(i).

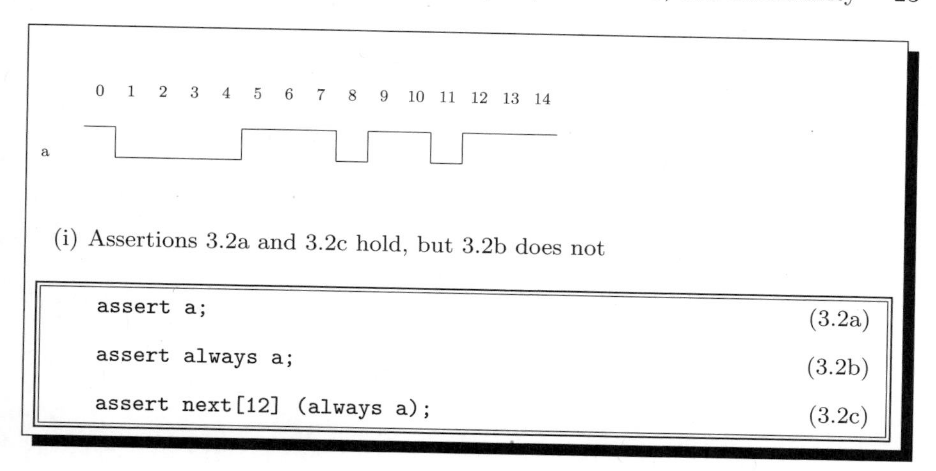

(i) Assertions 3.2a and 3.2c hold, but 3.2b does not

```
assert a;                               (3.2a)
assert always a;                        (3.2b)
assert next[12] (always a);             (3.2c)
```

Fig. 3.2: The importance of the `always` operator

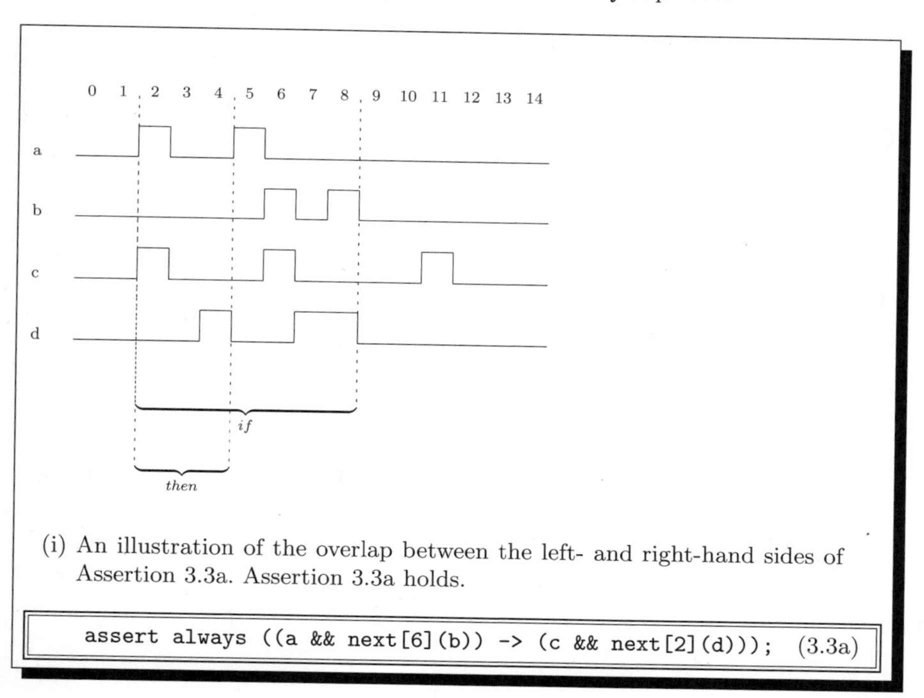

(i) An illustration of the overlap between the left- and right-hand sides of Assertion 3.3a. Assertion 3.3a holds.

```
assert always ((a && next[6](b)) -> (c && next[2](d)));   (3.3a)
```

Fig. 3.3: The current cycle

Assertion 3.3a holds on Trace 3.3(i) because the sub-property `(a && next[6] (b)) -> (c && next[2] (d))` holds at all cycles. It holds at cycle 2 because the left-hand side `(a && next[6] (b))` holds and in addition the right-hand side `(c && next[2] (d))` holds. It holds at all other cycles because if the "if" part of a logical implication does not hold, then the "else" part defaults to true.

Previously, for example in examining Assertions 2.2a and 2.3b, we have seen cases where there is an overlap between the cycles involved in satisfying two different occurrences of the left-hand side of a logical implication. These overlaps are caused by the presence of the `always` operator in the property. If we remove the always operator, the issue of overlap disappears from Assertions 2.2a and 2.3b. The overlap of Assertion 3.3a is different from the overlap of Assertions 2.2a and 2.3b in a very important way: Assertion 3.3a is written in such a way that the cycles involved in calculating the left-hand side of the logical implication overlap those involved in calculating the right-hand side of the logical implication: calculating the "if" part of the logical implication involves examining the current cycle and also "looking ahead" six cycles, while calculating the "then" part of the logical implication involves examining the current cycle and also "looking ahead" two cycles. Thus the overlap results from the logical implication itself. This style of PSL property is usually very confusing to new users of PSL, and indeed it is not an intuitive way to use the language. For this reason, such properties are not recommended, and the *simple subset* of PSL restricts the language in such a way that such properties are not allowed. We discuss the simple subset in more detail later. For now, remember that if you have written a property in which the cycles involved in calculating the left-hand side of an operator overlap for more than one cycle with those involved in calculating the right-hand side of the same operator, you have not written a property that is in the simple subset.

3.5 Reporting a failure

Consider Assertion 3.4a on the traces shown in Figure 3.4. Obviously, Assertion 3.4a does not hold on them. Now consider four different verification tools, each of which reports the failure of Assertion 3.4a as indicated by the signal called "failure" in Traces 3.4(i), 3.4(ii), 3.4(iii) and 3.4(iv). Tool 1 reports a failure at cycle 4, the earliest that it can be detected. Tool 2 reports a failure at cycle 4, but also at cycle 7, when `b` is asserted for a second time. Tool 3 reports a failure at the end of the trace, and Tool 4 reports a failure at cycle 1, when `a` is asserted.

Which tool is correct? The answer is that they all are. PSL defines whether or not a property holds on a trace – that is all. It says nothing about when a tool, dynamic or static, should report on the results of the analysis. Thus, there is no meaning in asking whether Tool 1, 2, 3 or 4 is correct. All of them indicate that the property fails on the trace, so all are correct. The

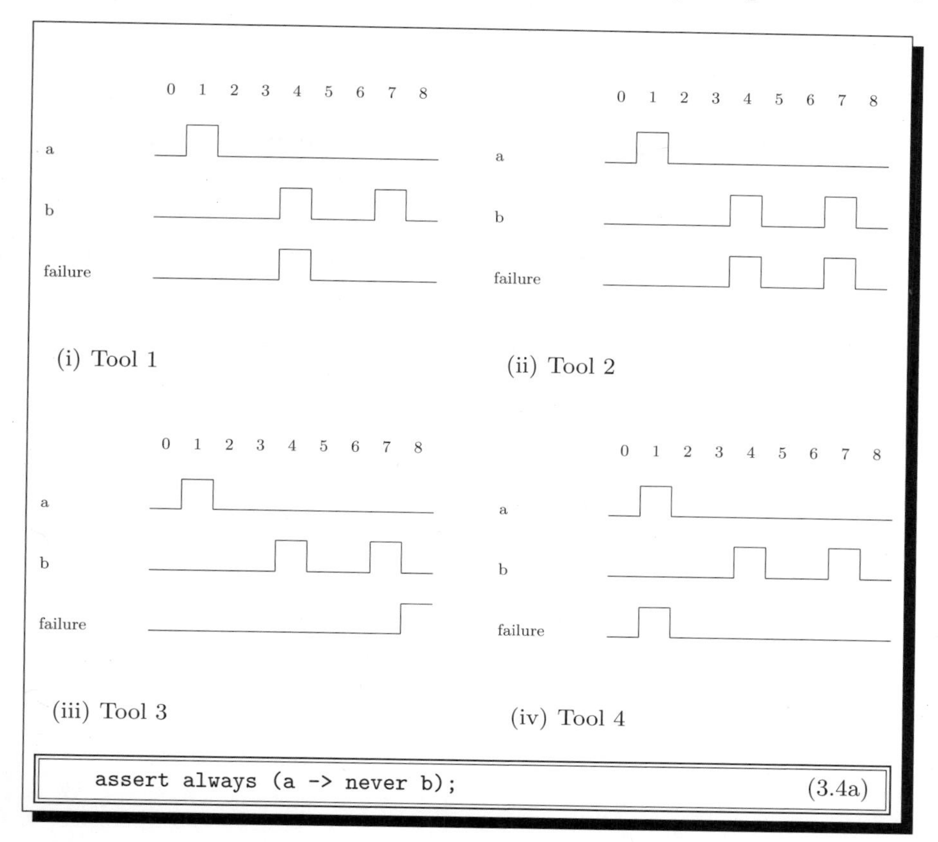

Fig. 3.4: Four different tools reporting the failure of Assertion 3.4a on the same trace

failure indications shown along the trace can be thought of as debugging aids, hinting to the user where to look along the trace for the failure. As far as PSL is concerned, as long as a tool indicates correctly whether or not a property holds on a trace, it has done its job.

4

Weak vs. Strong Temporal Operators

Temporal operators can be *weak* or *strong*. A strong temporal operator such as `eventually!` is indicated by an exclamation point (`!`) as part of its name. An operator without an exclamation point, such as `next`, is weak. Up until now we have seen only one version of each operator, but many of the operators we have seen previously come in both weak and strong versions. The difference between weak and strong operators is important when the path is "too short" to determine whether or not everything that needs to happen to satisfy a property has indeed happened.[1] For instance, consider the specification "every assertion of signal `a` must be followed three cycles later by an assertion of signal `b`", on Trace 4.1(i). Does the specification hold or not? The assertion of signal `a` at cycle 2 is satisfied by the assertion of signal `b` at cycle 5. But what about the assertion of signal `a` at cycle 9? It requires an assertion of signal `b` at cycle 12, but the trace ends before cycle 12 is reached. Weak and strong temporal operators allow us to distinguish between the case where we would like to say that our specification holds on Trace 4.1(i), and the case where we would like to say that it does not.

A weak temporal operator is lenient – in the case that a trace ends "too soon", a property using a weak temporal operator will hold as long as nothing else has gone wrong. If we were to continue the verification run shown in Trace 4.1(i) for a few more cycles, we might find out that signal `b` is asserted in cycle 12. It *could* happen. Thus, Assertion 4.1a, which uses the weak `next` operator, holds on Trace 4.1(i).

A strong temporal operator is strict – in the case that a trace ends "too soon", a property using a strong operator will not hold, even if nothing else has gone wrong. There is no way to know what will happen if we continue the verification run shown in Trace 4.1(i) for a few more cycles, and we shouldn't

---

[1] In this chapter we assume finite paths such as those seen in simulation. On infinite paths, such as those seen in formal verification, there is also a difference between weak and strong operators. We discuss this issue in Chapter 11.

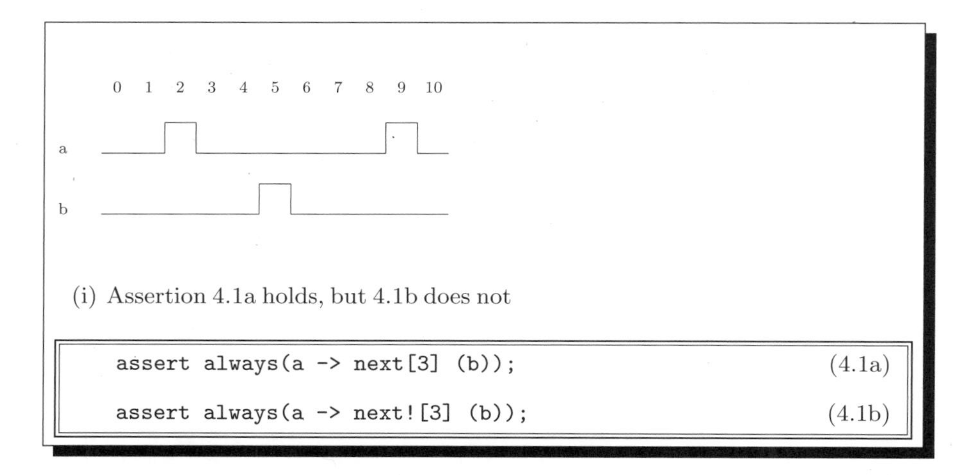

Fig. 4.1: A trace illustrating the idea behind weak and strong operators

say that a property holds without having all the information. Thus, Assertion 4.1b, which uses the strong `next!` operator, does not hold on Trace 4.1(i).

NOTE: A weak temporal operator is lenient even about `'false`, and a strong temporal operator is strict even about `'true`. Thus, Assertion 4.2a holds on Trace 4.2(i) because `'false` is treated as something that could happen in the future, and Assertion 4.2b does not hold on Trace 4.2(i) because `'true` is not treated as a sure thing.

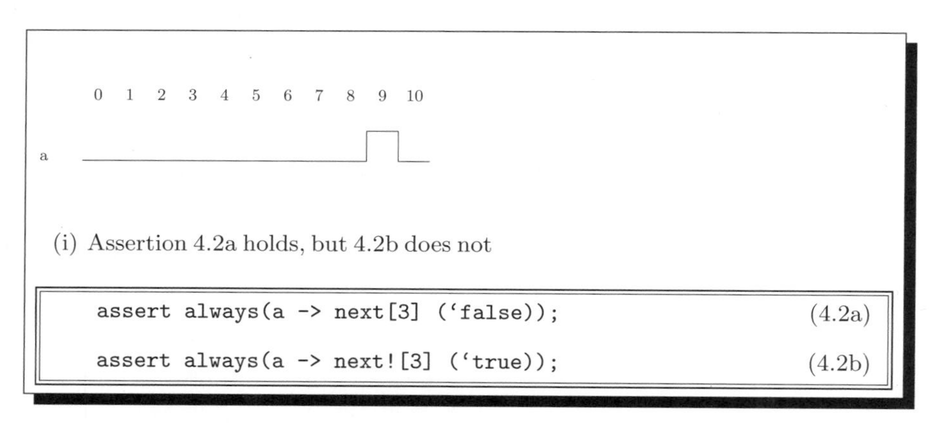

Fig. 4.2: A weak temporal operator is lenient even about `'false`, and a strong temporal operator is strict even about `'true`

Below we present the families of strong operators, in the same order in which their weak cousins were presented, and contrast the weak and strong versions.

4.1 The next! operator

Assertion 4.3a states that whenever a holds, then b should hold in the next cycle. Using the strong version of the next operator, we get Assertion 4.3b which requires that in addition, there *be* a next cycle. For example, while Assertion 4.3a holds on Trace 4.3(i), Assertion 4.3b does not, because even though b is asserted after the first two assertions of a, the trace ends too soon with regards to the third assertion of a.

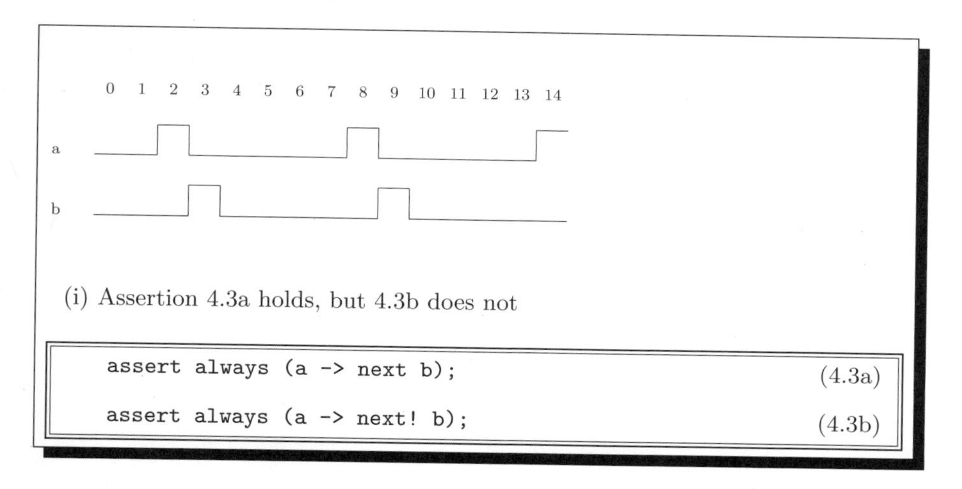

(i) Assertion 4.3a holds, but 4.3b does not

```
assert always (a -> next b);      (4.3a)

assert always (a -> next! b);     (4.3b)
```

Fig. 4.3: Weak vs. strong next

4.2 Variations on next! including next_event!

The strong next_event! operator is related to its weak version next_event in the same way that the strong next! operator is related to next. For example, Assertion 4.4b does not hold on Trace 4.4(i), because there are not enough grants. However, the weak version, Assertion 4.4a, does hold on the same trace.

The next![n] operator allows you to indicate that at least n next cycles are needed, and the next_event!(b)[n] operator allows you to indicate that at least n occurrences of the Boolean event b are needed.

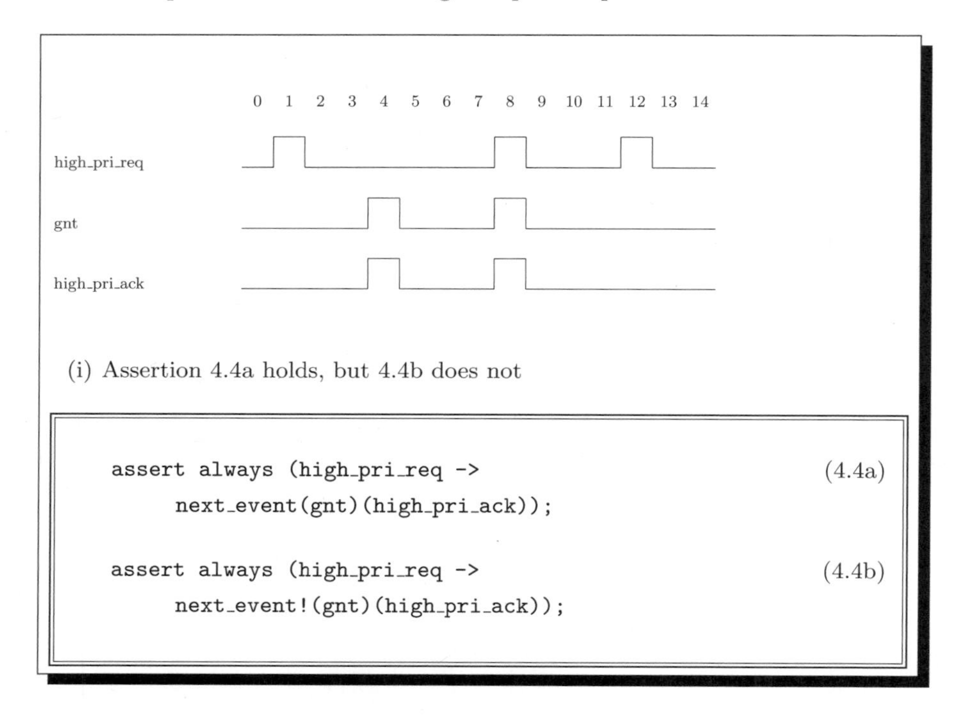

Fig. 4.4: Weak vs. strong `next_event`

You should by now be able to guess the meaning of the `next_a![i:j]` and `next_event_a!(b)[i:j]` operators. The `next_a![i:j]` operator holds if there are at least j additional cycles, and its operand holds on all of the ith through j^{th} of them, inclusive. The `next_event_a!(b)[i:j]` operator holds if there are at least j additional cycles on which `b` holds, and the second operand holds on all of the ith through j^{th} of them, inclusive.

The `next_e![i:j]` operator creates a property that holds if its operand holds on at least one of the ith through j^{th} next cycles, inclusive. There do not have to be j cycles if the operand holds on some cycle between the ith and the jth, inclusive. Similarly, the `next_event_e!(b)[i:j]` operator creates a property that holds if its second operand holds on at least one of the ith through j^{th} next occurrences of `b`, inclusive. There do not have to be j occurrences if the second operand holds on some occurrence between the ith and the jth, inclusive. For example, consider Assertion 4.5b, which states that a request (assertion of `req`) must be acknowledged (assertion of `ack`) on one of the next four grants, and that in addition, the trace must not end before there is an appropriate grant. Assertion 4.5b does not hold on Trace 4.5(i) because none of the three grants shown is acknowledged. However, it does hold on Trace 4.5(ii) because at least one of the first, second, third or fourth grants is acknowledged. The fact that the fourth grant does not occur is immaterial

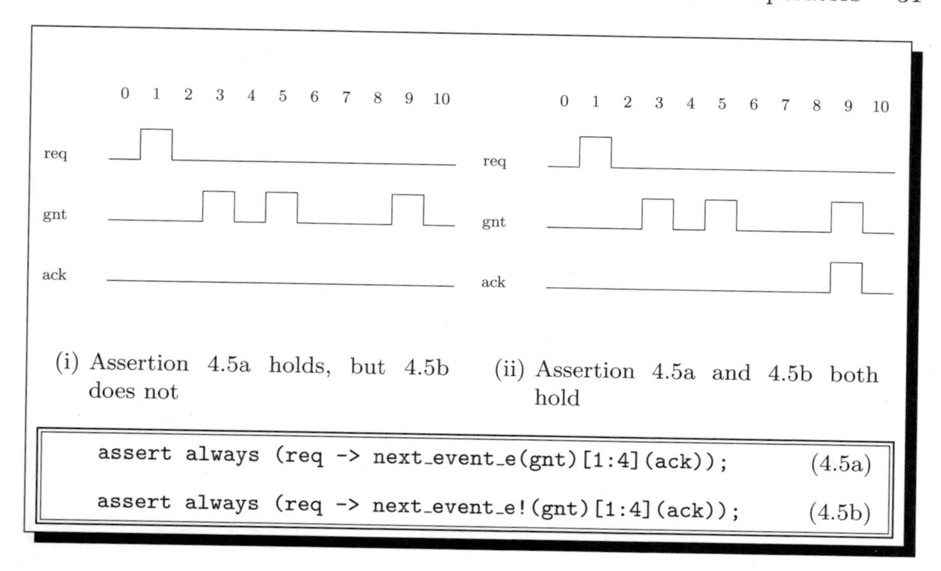

Fig. 4.5: Weak vs. strong `next_event_e[i:j]`

– there is no way for a fourth grant to change the fact that we know for sure that one of the first, second, third or fourth grants is acknowledged.

The weak version of Assertion 4.5b holds on both Trace 4.5(i) and Trace 4.5(ii). The only way to violate the weak version shown in Assertion 4.5a is for there to be four grants, none of which are acknowledged.

4.3 The `until!` and `until!_` operators

We have previously seen assertions like Assertion 4.6a, which states that whenever `a` is asserted, then starting the next cycle `b` should be asserted until `c` is asserted. For example, Assertion 4.6a holds on Trace 4.6(i). But what about a trace like that shown in Trace 4.6(ii), where `c` never arrives? Does Assertion 4.6a hold on Trace 4.6(ii) or not? It does. The reason is that the `until` operator is a weak operator.

If we want to express the requirement of Assertion 4.6a, but in addition insist that `c` eventually occur, we need to use a strong operator. Assertion 4.6b, which uses a strong operator, does not hold on Trace 4.6(ii).

The `until!_` operator is the strong version of the `until_` operator. The `until!_` operator holds only if its left operand stays asserted up to and *including* the cycle where its right operand is asserted, and in addition its right operand eventually holds.

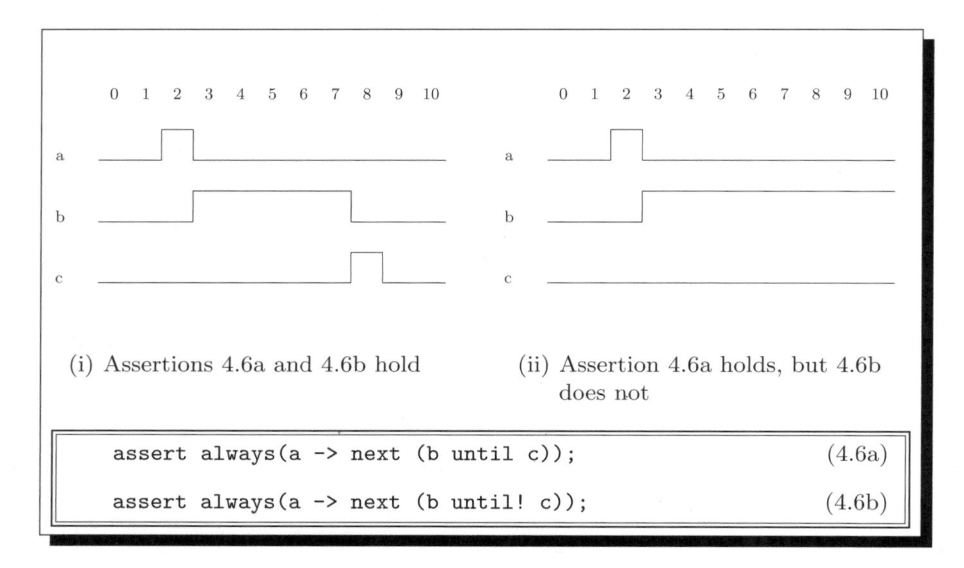

Fig. 4.6: Weak vs. strong `until`

4.4 The `before!` and `before!_` operators

The emphasis of the strong versions of the `before` operators is on their left-hand operand. For example, Assertion 4.7a states that following an assertion of `req`, an assertion of `gnt` is required before `req` can be asserted again, but

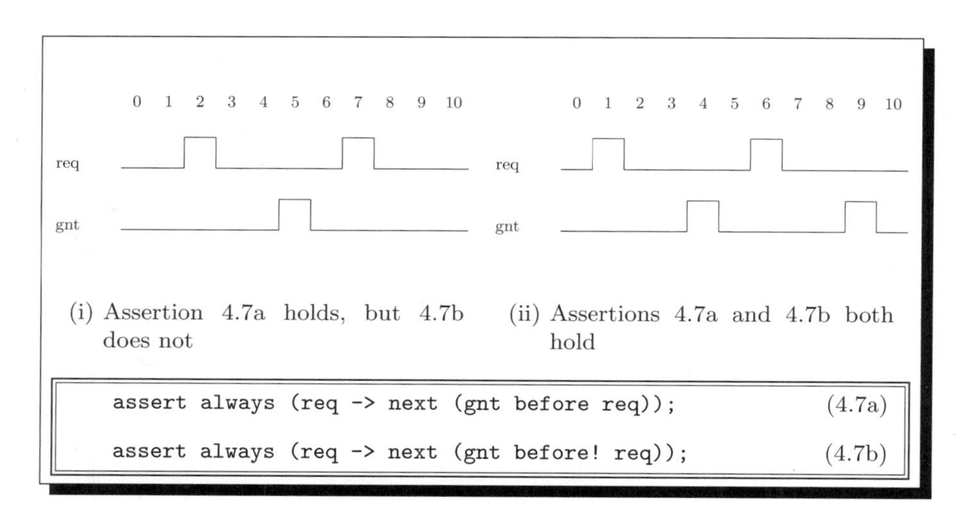

Fig. 4.7: Weak vs. strong `before`

being weak holds not only on Trace 4.7(ii), but also on Trace 4.7(i), in which after the second request neither a grant nor a third request ever arrives. The strong version, shown in Assertion 4.7b, requires that a grant eventually be given, but not necessarily that another request be seen. Thus, Assertion 4.7b does not hold on Trace 4.7(i), but holds on Trace 4.7(ii).

The `before!_` operator is the strong version of the `before_` operator. Thus, Assertion 4.8a holds on Trace 4.8(i), in which the first request (assertion of signal `req`) is granted (assertion of signal `gnt`) but the second is not. The strong version of Assertion 4.8a, shown in Assertion 4.8b, does not hold on Trace 4.8(i) because the second request is not granted. Both Assertion 4.8a and 4.8b hold on Trace 4.8(ii).

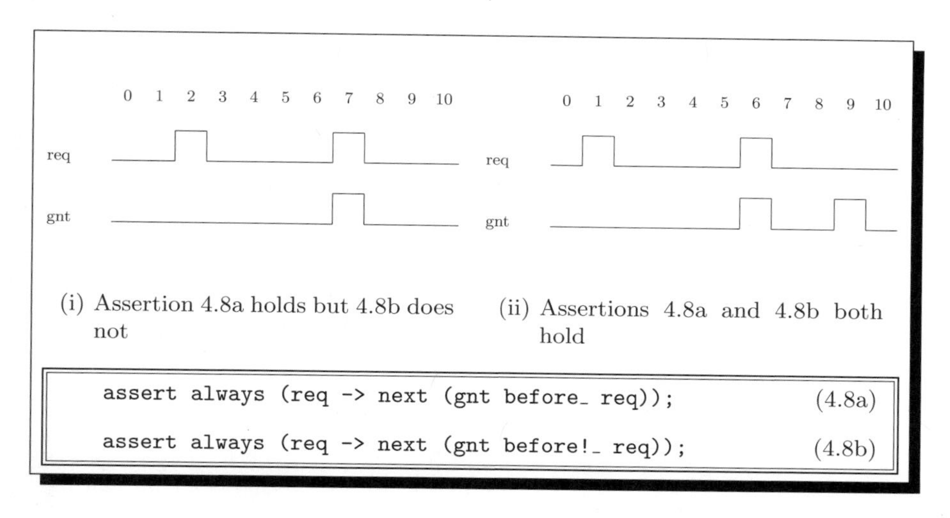

Fig. 4.8: Weak vs. strong `before_`

4.5 Operators without weak or strong versions

We have seen weak and strong versions of many operators. The `eventually!` operator is not a strong version of some other operator. In other words, there is no weak version of `eventually!`. The reason is that the meaning of other strong operators can be broadly described as requiring that nothing bad must happen up until some terminating condition, and in addition, the terminating condition must eventually occur. The weak versions then waive the requirement that the terminating condition eventually occur, leaving only the requirement that nothing bad must happen. The `eventually!` operator, however, contains no idea of a bad "something", only a terminating condition.

Therefore, there is no way to weaken it without emptying it completely of meaning.

Just as there is no weak version of `eventually!`, there is no strong version of the operators `always` and `never`, for the opposite reason. In the case of `always` and `never`, there is no terminating condition to waive.

5

SERE Style

Up until now, we have seen PSL properties that are built of Boolean expressions and temporal operators in LTL style. Another way to build properties is to use SEREs – Sequential Extended Regular Expressions. SEREs are similar in spirit to standard regular expressions, like those used for pattern matching in many applications. One difference is that the atoms of a SERE are Boolean expressions, whereas the atoms of a standard regular expression are single characters.

A SERE is typically enclosed in curly braces, and the atoms of the SERE are typically separated by semi-colons. For instance, {`a;b;c`} is a SERE, and SERE 5.1a is a more complicated SERE. SERE 5.1a describes a sequence of cycles in which `(req_out && !ack)` is asserted on the first cycle, then `(busy && !ack)` is asserted for zero or more cycles, indicated by the consecutive repetition operator `[*]`, and finally `ack` is asserted. Thus, SERE 5.1a matches the sequence of cycles labeled as "1" in Trace 5.1(i).

Don't be tempted into reading more into a SERE than is actually there: SERE 5.1a matches the sequence of cycles labeled as "2" in Trace 5.1(i) as well. SERE 5.1a does not prohibit the assertion of `busy` during the last cycle of the SERE. If it is important to exclude such behavior, `busy` must be mentioned explicitly, as shown in SERE 5.1b. SERE 5.1b matches the sequence of cycles labeled as "1" shown in Trace 5.1(i), but does not match the sequence of cycles labeled as "2" in that trace.

How is a SERE used? First, since a SERE is a property, it can be used as a sub-property. For example, Property 5.2a holds if whenever there is an assertion of `req_in && gnt` then starting on the next cycle we see a sequence of cycles matching SERE 5.1a. Property 5.2a holds on Trace 5.2(i).

NOTE: A SERE is a property, but a property is not a SERE. Thus, while you can use a SERE as an operand of a temporal operator, you cannot embed a temporal operator such as `always` or `next` inside of a SERE.

Suppose now that we have a situation similar to that of Property 5.2a, but in which a grant (assertion of signal `gnt`) is given the cycle *after* the assertion of `req_in`. We could try to replace the Boolean expression `req_in && gnt` with

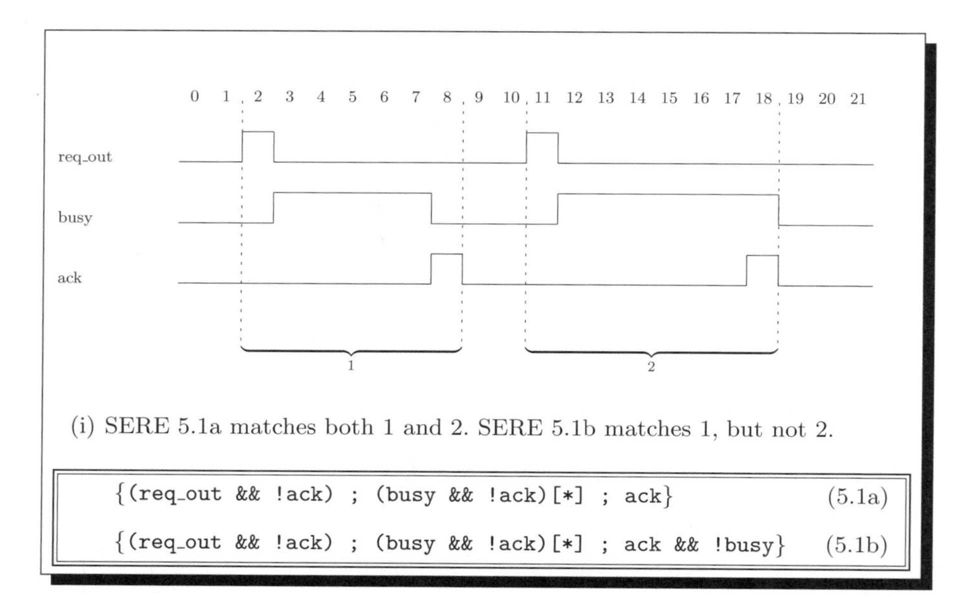

Fig. 5.1: Two simple SEREs

the temporal expression `req_in && next gnt`, as in Property 5.3a. However, remember the lesson of Section 3.4: in Property 5.3a the current cycle of the left-hand side (`req_in && next gnt`) is the same as the current cycle of the right-hand side (because they are connected by the Boolean operator `->`). Thus, the current cycle of `gnt` (which is the operand of a `next` operator) is the same as the current cycle of the SERE (which is also the operand of a `next` operator). This is slightly confusing, and indeed Property 5.3a is not in the *simple subset* of PSL discussed in Chapter 9.

Let's modify Property 5.3a as shown in Property 5.3b. Now a single `next` operator is applied to both `gnt` and the SERE, which are both operands of the `->` operator. Property 5.3b is equivalent to Property 5.3a, but is in the simple subset, making the timing between `gnt` and the SERE easier to see. If we mean that the current cycle of the SERE should be the cycle *after* that of `gnt`, we can manipulate Property 5.3b by adding a `next` as in Property 5.3c. However, the suffix implication operators provide a much easier way.

5.1 Suffix implication (`|->` and `|=>`)

The *suffix implication* operators (`|->` and `|=>`) provide a way to link two SEREs in such a way that the right-hand side starts when the left-hand side finishes. The overlapping suffix implication operator (`|->`) interprets "when the left-hand side finishes" as "at the same cycle as the cycle in which the

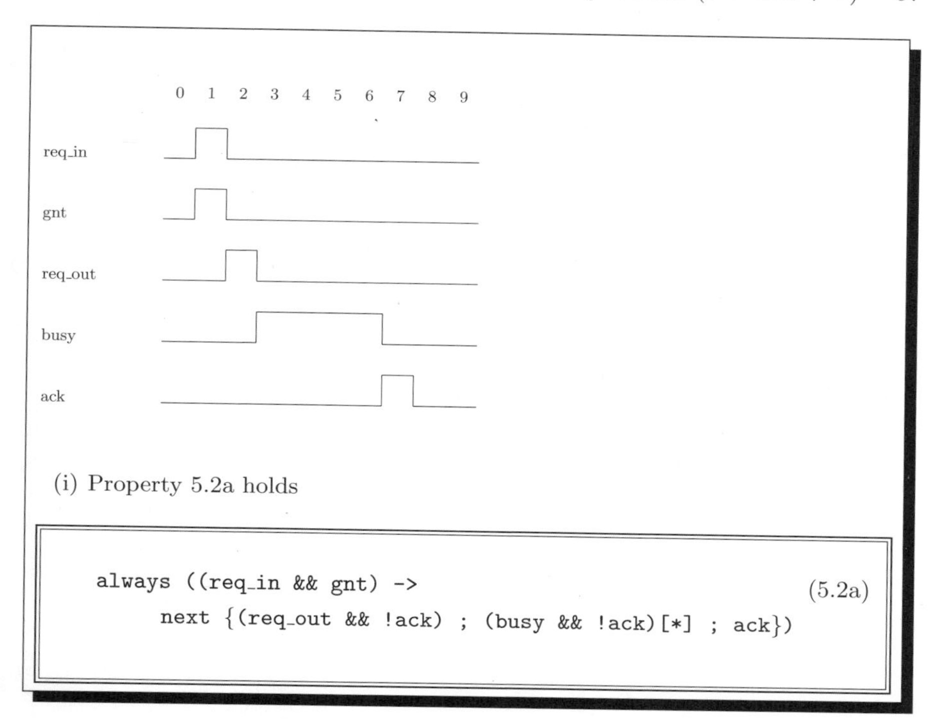

Fig. 5.2: A SERE is a property

```
always ((req_in && next gnt) ->                                    (5.3a)
      next {(req_out && !ack) ; (busy && !ack)[*] ; ack})

always (req_in -> next (gnt ->                                     (5.3b)
      {(req_out && !ack) ; (busy && !ack)[*] ; ack}))

always (req_in -> next (gnt ->                                     (5.3c)
      next {(req_out && !ack) ; (busy && !ack)[*] ; ack}))
```

Fig. 5.3: Property 5.3a is not in the simple subset. Properties 5.3b and 5.3c are in the simple subset, but are difficult to read. These properties can be expressed more easily using suffix implication.

left-hand side finishes", while the non-overlapping suffix implication operator (|=>) interprets it as "the cycle after the cycle in which the left-hand side finishes". Thus, Property 5.4a is equivalent to Property 5.3b, and Property 5.4b is equivalent to Property 5.3c. Both Property 5.4a and Property 5.4b are easier to grasp than the equivalent property without suffix implication, and both belong to the simple subset, discussed in Chapter 9.

Recall that the logical implication operator (->) can be understood as an if-then expression, with the else-part being implicitly true. The suffix implica-

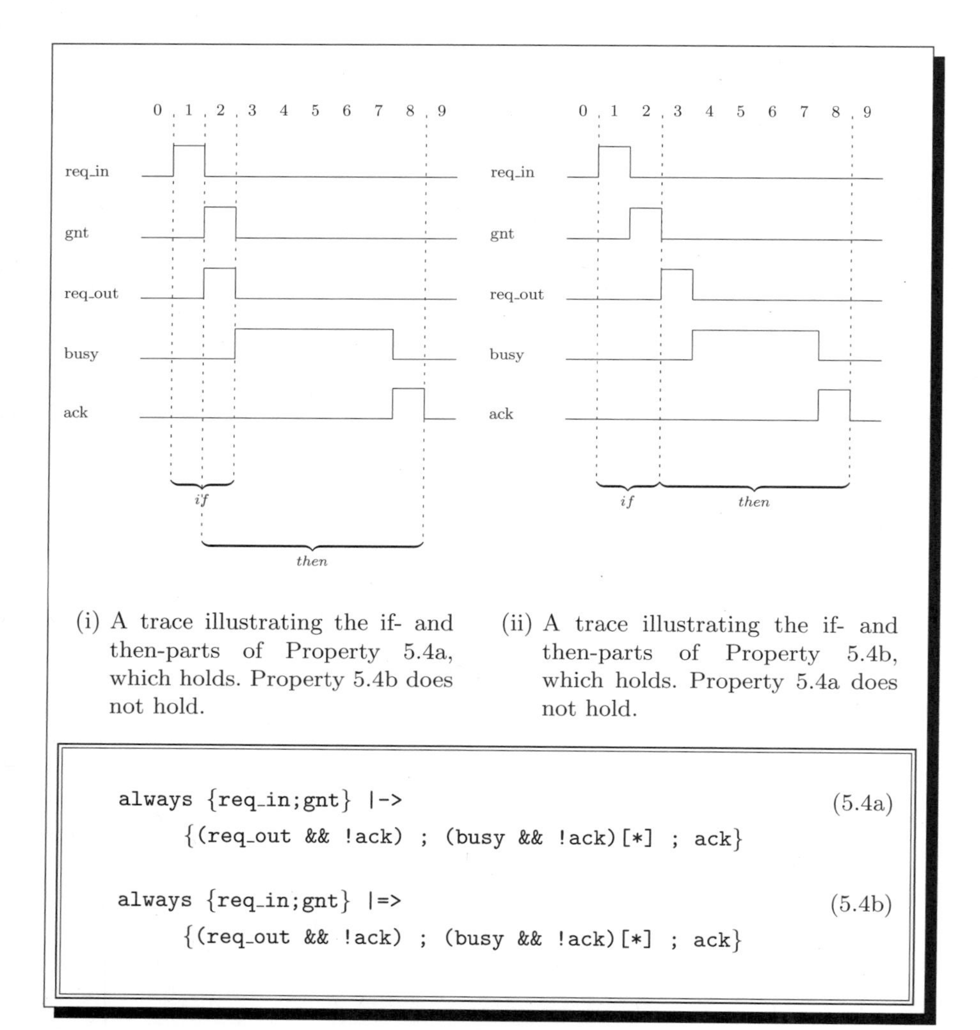

(i) A trace illustrating the if- and then-parts of Property 5.4a, which holds. Property 5.4b does not hold.

(ii) A trace illustrating the if- and then-parts of Property 5.4b, which holds. Property 5.4a does not hold.

```
always {req_in;gnt} |->                                        (5.4a)
      {(req_out && !ack) ; (busy && !ack)[*] ; ack}

always {req_in;gnt} |=>                                        (5.4b)
      {(req_out && !ack) ; (busy && !ack)[*] ; ack}
```

Fig. 5.4: Suffix implication

tion operators (|-> and |=>) can be understood the same way. The difference between the logical implication operator (->) and the suffix implication operators (|-> and |=>) is in the timing relationship between the if- and the then-parts. While the current cycle of the then-part of a logical implication operator (->) is the same as the current cycle of its if-part, the current cycle of the then-part of a suffix implication operator (|-> or |=>) is the first cycle of the *suffix* of the trace that remains once the if-part has been seen. This is illustrated for Properties 5.4a and 5.4b in Traces 5.4(i) and 5.4(ii).

In Traces 5.4(i) and 5.4(ii) there is a single assertion of signal `req_in`. If there are multiple assertions of `req_in`, then of course we will be able to identify multiple if-then pairs, as shown in Trace 5.5(i). Note that the if-part of

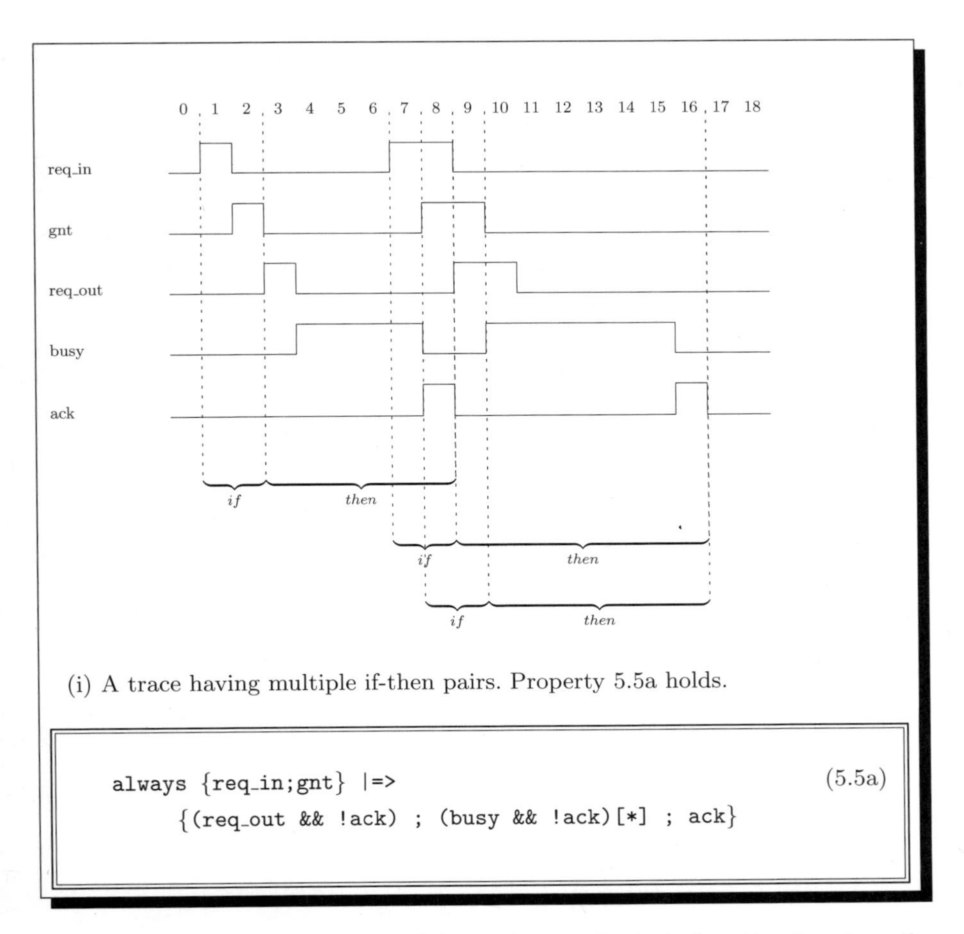

(i) A trace having multiple if-then pairs. Property 5.5a holds.

```
always {req_in;gnt} |=>
    {(req_out && !ack) ; (busy && !ack)[*] ; ack}        (5.5a)
```

Fig. 5.5: The if-part of another if-then pair may begin before the if-part or the then-part of a previous if-then pair has completed

another if-then pair may begin before the if-part or the then-part of a previous if-then pair has completed. For instance, the second if-then pair starts at cycle 7, before the then-part of the first if-then pair has completed. The third if-then pair starts at cycle 8, before the then-part of the first if-then pair has completed, and before the if-part of the second if-then pair has completed. We have seen this kind of overlapping previously, in Traces 2.2(ii) and 2.3(iii). Note that in Trace 5.5(i), the assertion of `ack` at cycle 16 completes the then-part of the second and of the third if-then pair. For a deeper discussion of this phenomenon, see Section 13.4.2.

Properties 5.6a and 5.6b hold on Trace 5.6(i) because there is no sequence of cycles matching the left-hand side, thus the else-parts default to true at every cycle.

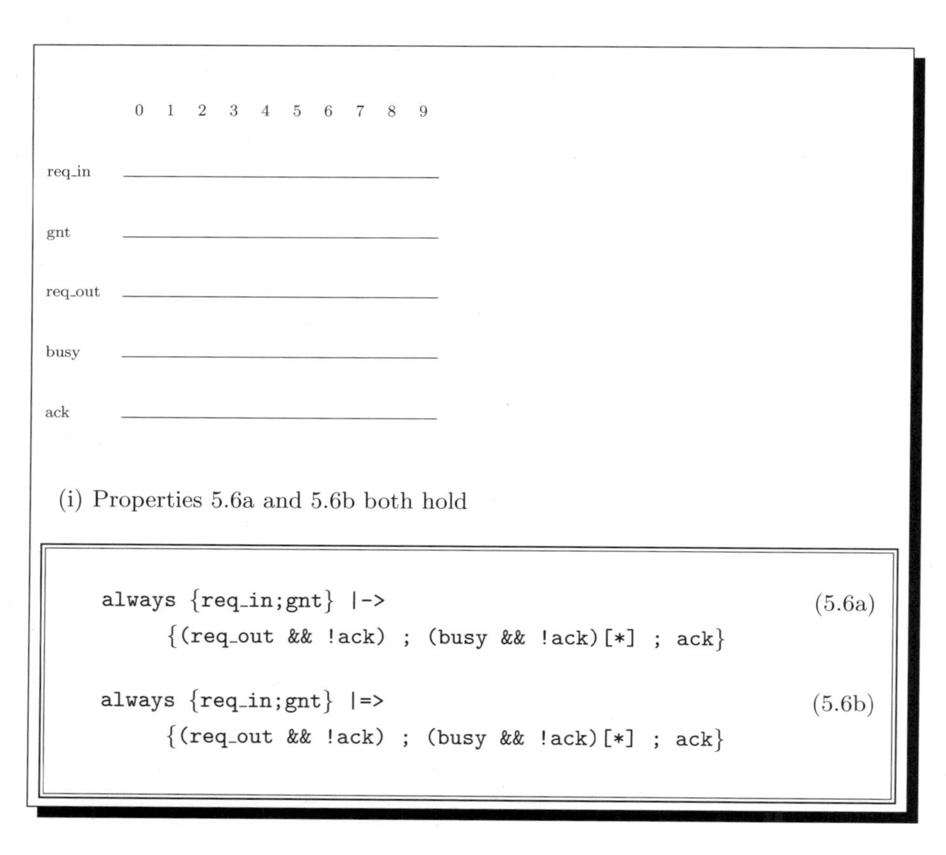

Fig. 5.6: The else-part of a suffix implication defaults to true

5.2 Weak and strong SEREs

Like temporal operators, SEREs come in weak and strong versions, and like temporal operators, the strong version of a SERE is indicated by an exclamation point (`!`). Thus, Property 5.7a, whose right-hand side is a weak SERE, holds even if signal `ack` is never asserted (as long as the rest of the SERE is not violated). Property 5.7b, whose right-hand side is a strong SERE, holds only if we eventually reach a cycle where `ack` occurs. In other words, the weak version of a SERE allows us to accept a trace that is "too short", whereas the strong version requires that we "reach the end" of the SERE. Thus, while Property 5.7a holds on Trace 5.7(i) as well as Trace 5.7(ii), Property 5.7b, the strong version, holds on Trace 5.7(ii) but not on Trace 5.7(i).

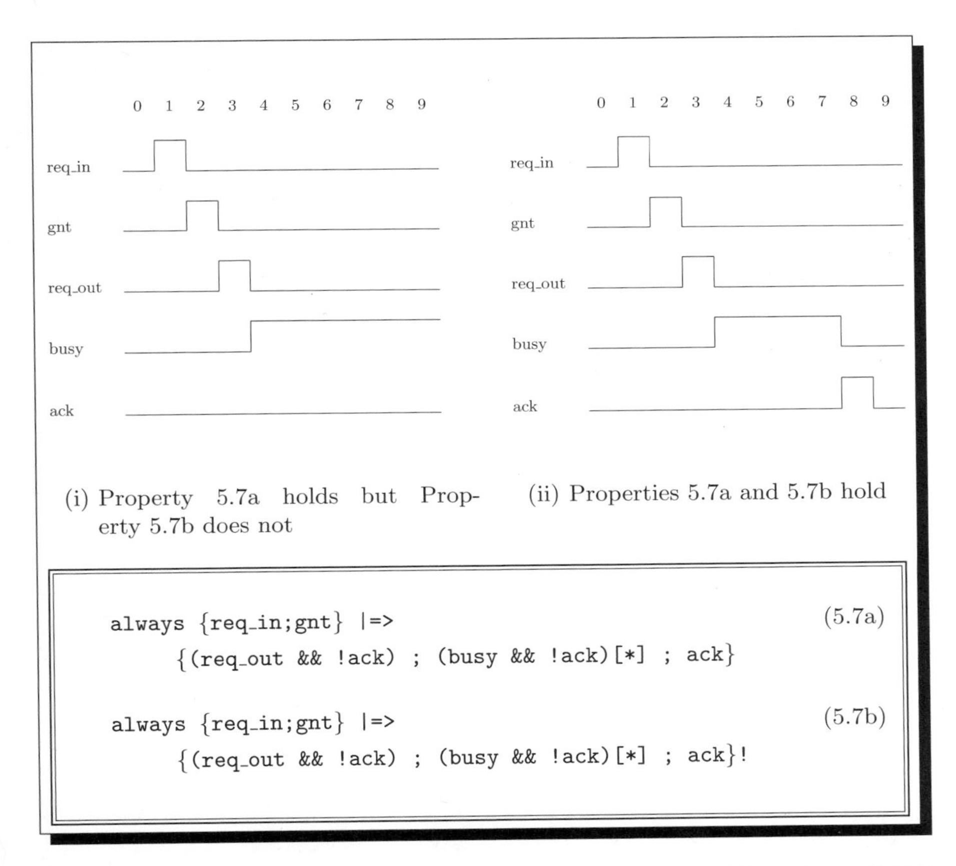

Fig. 5.7: A weak SERE holds even if the trace is "too short", while a strong SERE must "reach the end"

5.3 The never operator applied to a SERE

Another way to use a SERE is in describing sequences of events that should never happen. For example, Assertion 5.8a states that signal `req` should never be asserted for two consecutive cycles, and thus does not hold on Trace 5.8(i).

As another example, consider Assertion 5.9a. It states that an acknowledged high priority request (assertion of `req` together with `high_pri`, followed

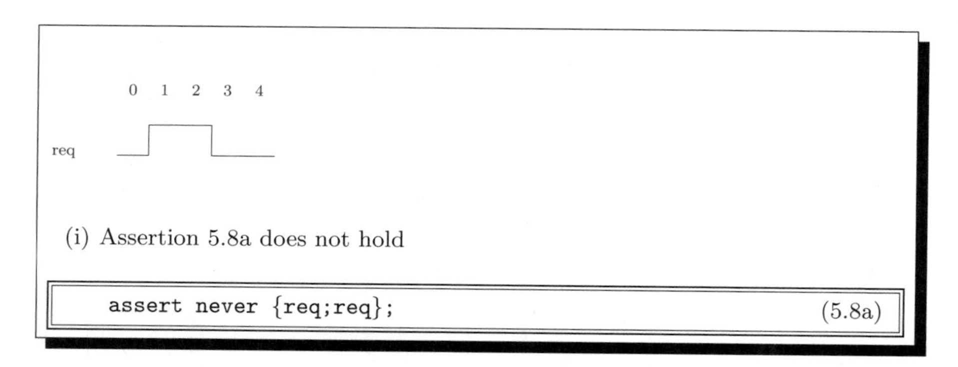

Fig. 5.8: The never operator

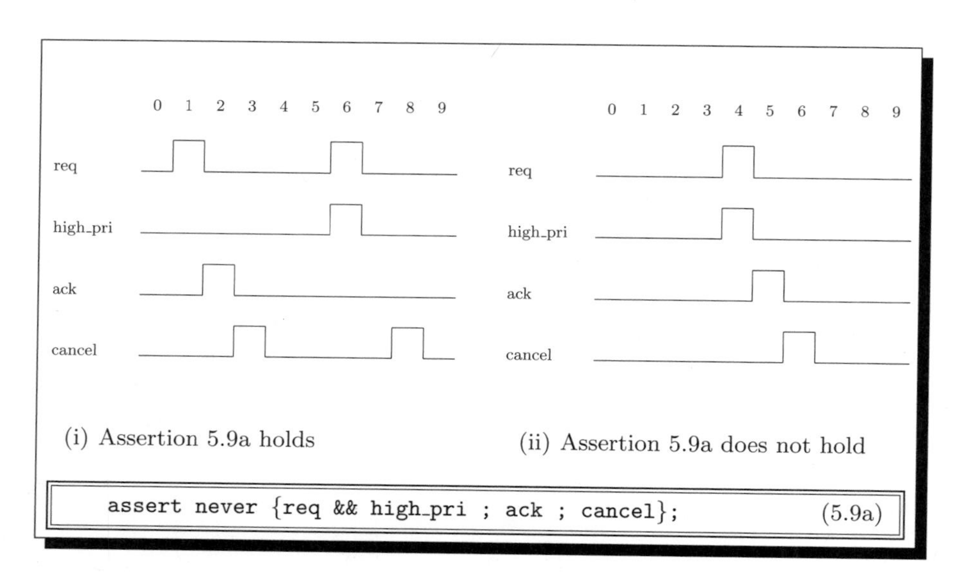

Fig. 5.9: More about the never operator

a cycle later by ack) cannot be canceled (assertion of cancel the cycle after ack). Thus it holds on Trace 5.9(i), but not on Trace 5.9(ii).

5.4 SERE repetition operators ([*n], [=n], and [->n])

Up until now, we have seen basic SEREs composed of (possibly repeated) Boolean expressions separated by semi-colons. In this section, we examine SERE operators that allow you to build more sophisticated SEREs, using variations on the SERE repetition operators [*n], [=n], and [->n].

Consecutive repetition operators provide a shortcut to typing the same sub-SERE a number of times. The [*n] operator is an abbreviation for n repetitions of the SERE it is applied to. For example, instead of typing busy;busy;busy in Assertion 5.10a, we can use the abbreviation busy[*3], as in Assertion 5.10b. Assertions 5.10a and 5.10b state that after a request (assertion of signal req), we expect to see an acknowledge (assertion of signal ack) followed by three cycles in which signal busy is asserted, followed by an assertion of signal done. Trace 5.10(i) is an example of a trace on which Assertions 5.10a and 5.10b hold.

Instead of a specific number of repetitions, we can specify a range, i through j, like this: [*i:j]. Thus, Assertion 5.10c is similar to Asser-

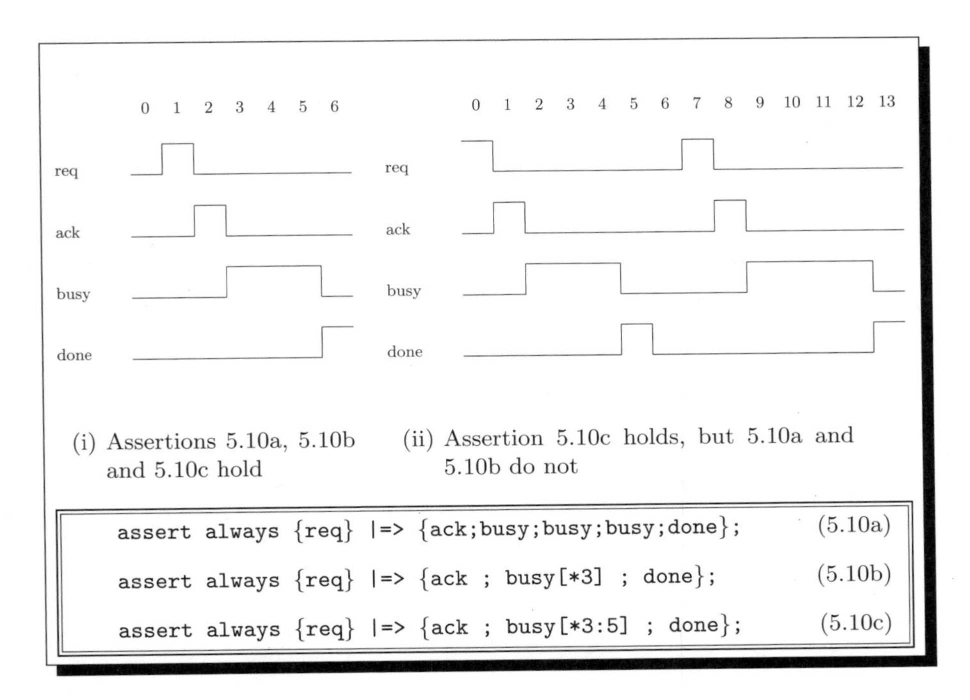

(i) Assertions 5.10a, 5.10b and 5.10c hold

(ii) Assertion 5.10c holds, but 5.10a and 5.10b do not

```
assert always {req} |=> {ack;busy;busy;busy;done};        (5.10a)

assert always {req} |=> {ack ; busy[*3] ; done};          (5.10b)

assert always {req} |=> {ack ; busy[*3:5] ; done};        (5.10c)
```

Fig. 5.10: The [*n] and [*i:j] operators

tion 5.10b, but instead of requiring exactly three assertions of signal `busy`, we require between three and five. Assertion 5.10c holds on Trace 5.10(i) as well as on Trace 5.10(ii).

The upper bound of a range can be infinity, which is indicated by a `$` in the SystemVerilog flavor, by *null* in the GDL flavor, and by `inf` in the Verilog, VHDL, and SystemC flavors. For instance, the SERE {`b[*5:inf]`} matches five or more repetitions of `b`.

We can repeat not only a Boolean expression, but a SERE as well. For example, Property 5.11a holds if after an acknowledged request (an assertion of signal `req` followed by an assertion of signal `ack`), we see three data transfers followed by an assertion of signal `d_end`, where a data transfer is an assertion of signal `start` followed by an assertion of signal `busy` followed by an assertion of signal `done`. Trace 5.11(i) is an example of a trace on which Property 5.11a holds.

If we omit the `n`, the resulting SERE matches any number of repetitions of the Boolean expression or SERE that is being repeated. For example, asserting Property 5.11b indicates that after an acknowledged request, we expect to see any number of data transfers followed by an assertion of `done`. Property 5.11b holds on Trace 5.11(i), as well as on Traces 5.11(ii) and 5.11(iii).

A `[*]` repetition matches any number of repetitions, including none. Thus, Property 5.11b holds on Trace 5.11(iv) as well. If you want to specify any non-zero number of repetitions, use the `[+]`. Thus, Property 5.11c is similar to Property 5.11b, but it requires at least one data transfer before signal `done` is asserted. Thus, Property 5.11c does not hold on Trace 5.11(iv), but it does hold on Traces 5.11(i), 5.11(ii) and 5.11(iii).

Instead of omitting the `n`, we can omit the Boolean expression or SERE and let the repetition `[*n]` stand alone. A stand-alone `[*n]` is equivalent to `'true[*n]`. In other words, `[*n]` matches any `n` cycles (because `'true` holds in all cycles). Another way to think of it is that it allows us to "skip" `n` cycles. Thus, Property 5.12a is similar to Property 5.11b, but instead of the acknowledge (assertion of signal `ack`) coming immediately following the request (assertion of signal `req`), it comes four cycles later. Property 5.12a holds on Trace 5.12(i).

The nonconsecutive repetition operator (`[=n]`) provides a way to describe repetitions that happen on not necessarily consecutive cycles. It can be applied only to a Boolean expression. For example, to describe the requirement that after a request (assertion of signal `req`), we expect to see an acknowledge (assertion of signal `ack`) followed by some number of cycles including three not necessarily consecutive assertions of signal `busy`, followed by an assertion of signal `done`, we can code Assertion 5.13a. Assertion 5.13a holds on Trace 5.13(i). Note the cycles marked "`busy[=3]`" in Trace 5.13(i). They do not start with an assertion of `busy`, nor do they end with one. The nonconsecutive repetition operator `[=n]` will match any sequence of cycles in which there are `n` not necessarily consecutive repetitions of the Boolean expression being repeated, including sequences of cycles in which the "padding" is at

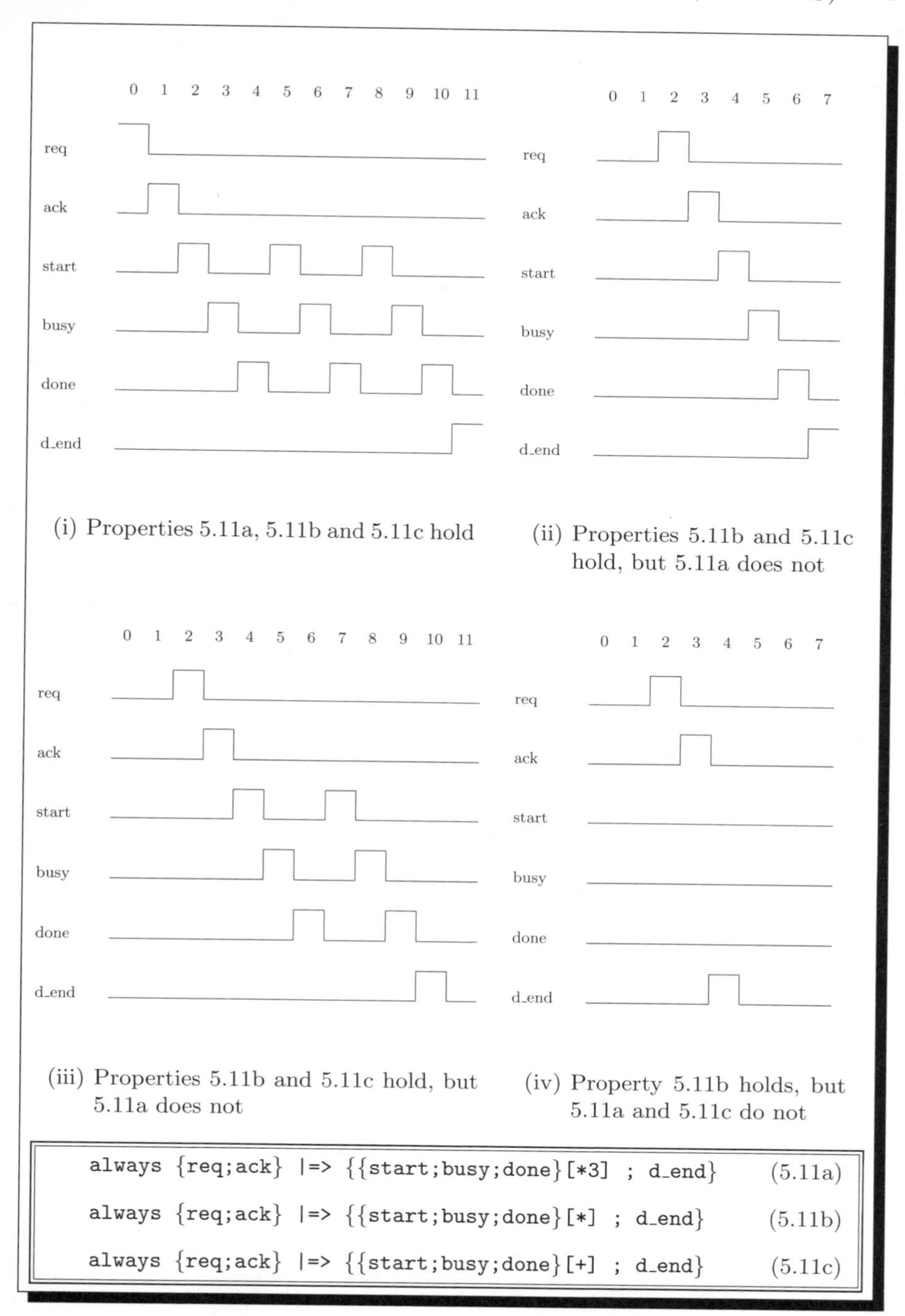

Fig. 5.11: Repeating a SERE

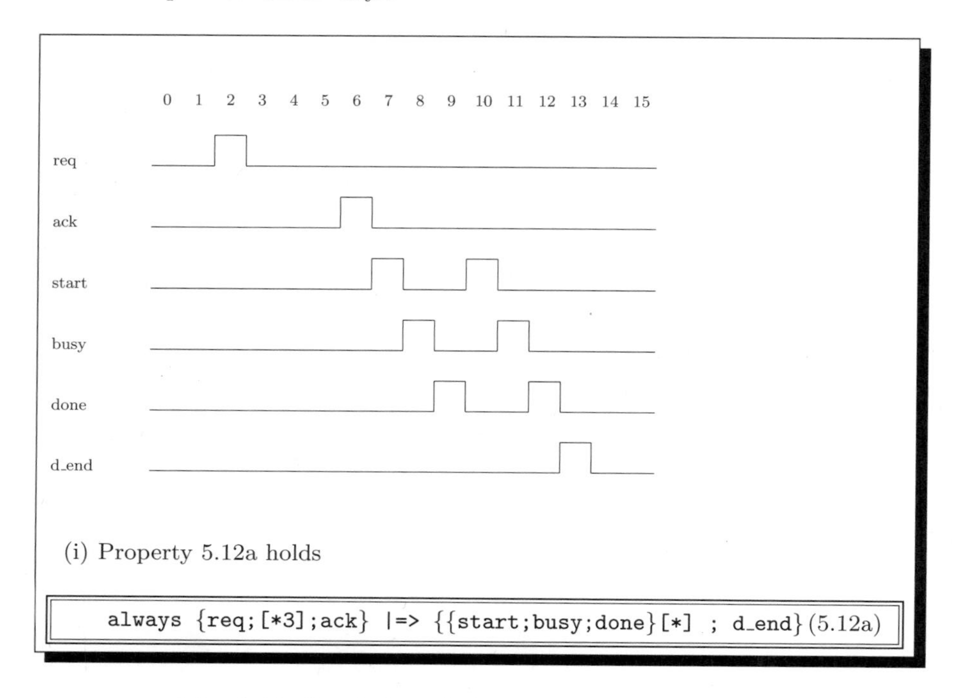

Fig. 5.12: Stand-alone `[*n]` is equivalent to `'true[*n]`

the beginning and/or at the end. In other words, it will match any sequence of cycles in which the number of assertions of the Boolean expression being repeated is *equal to* `n` (thus the use of the equals sign in `[=n]`).

If you want to disallow the padding at the end, use the goto repetition operator `[->n]`. The goto repetition operator `[->n]` is similar to the nonconsecutive repetition operator, except that the sequence of cycles being described ends with an assertion of the Boolean expression being repeated. In other words, it will match any sequence of cycles starting at the current cycle and ending after you "go to" the nth occurrence of the Boolean expression (the use of the `->` is intended to remind you of an arrow instructing you to go to `n`). Thus Assertion 5.13b does not hold on Trace 5.13(i), but it does hold on Trace 5.13(ii) because the third busy is immediately followed by an assertion of signal `done`.

The `n` can be omitted for the goto repetition operator, in which case `n` defaults to 1. In other words, `b[->]` is equivalent to `b[->1]`.

Like the consecutive repetition operators, the nonconsecutive repetition operator and the goto repetition operator can take a range. Thus, Assertion 5.14a requires three to five not necessarily consecutive assertions of busy after the assertion of signal `ack` and before the assertion of `done`, while Assertion 5.14b requires the same, and in addition that the assertion of signal `done` occur immediately after the 3rd, 4th, or 5th assertion of signal `busy`. Thus As-

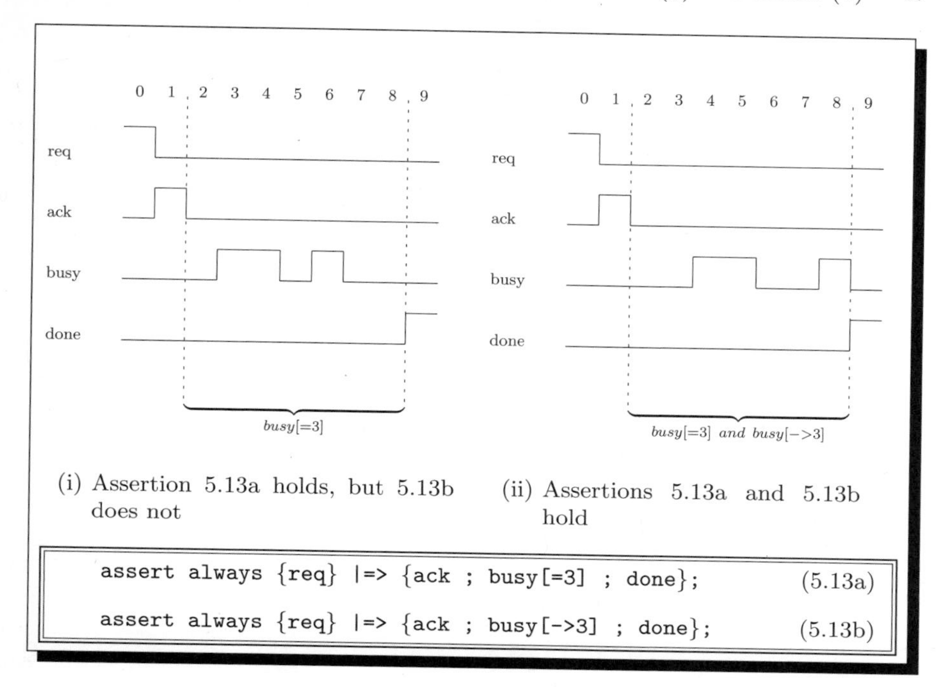


```
assert always {req} |=> {ack ; busy[=3] ; done};        (5.13a)
assert always {req} |=> {ack ; busy[->3] ; done};       (5.13b)
```

Fig. 5.13: Nonconsecutive repetitions using `[=n]` and `[->n]`

sertion 5.14a holds on Traces 5.14(i) and 5.14(ii), while Assertion 5.14b holds on Trace 5.14(ii) but not on Trace 5.14(i).

5.5 Concatenation (;) and fusion (:)

We have already seen the concatenation operator, a semi-colon, which is used to join two SEREs (or two Boolean expressions, or a Boolean expression and a SERE) in such a way that the right-hand SERE starts the cycle after the left-hand SERE ends. The fusion operator, a colon, is used to join two SEREs (or two Boolean expressions, or a Boolean expression and a SERE) in such a way that there is a single cycle of overlap between them: the right-hand SERE starts the same cycle that the left-hand SERE ends.

For example, consider the case that we want to specify the behavior of Assertion 5.14b, and in addition, following the assertion of signal `done`, we should see a data transfer, which consists of the assertion of signal `data` for some number of cycles (might be zero), followed by an assertion of signal `d_end`. Using the concatenation operator, we can write Assertion 5.15a. Trace 5.15(i) is a trace on which Assertion 5.15a holds. The first assertion of signal `req` gets three cycles of data before the assertion of `d_end`, while the second assertion of signal `req` sees four cycles of `data` before seeing `d_end`.

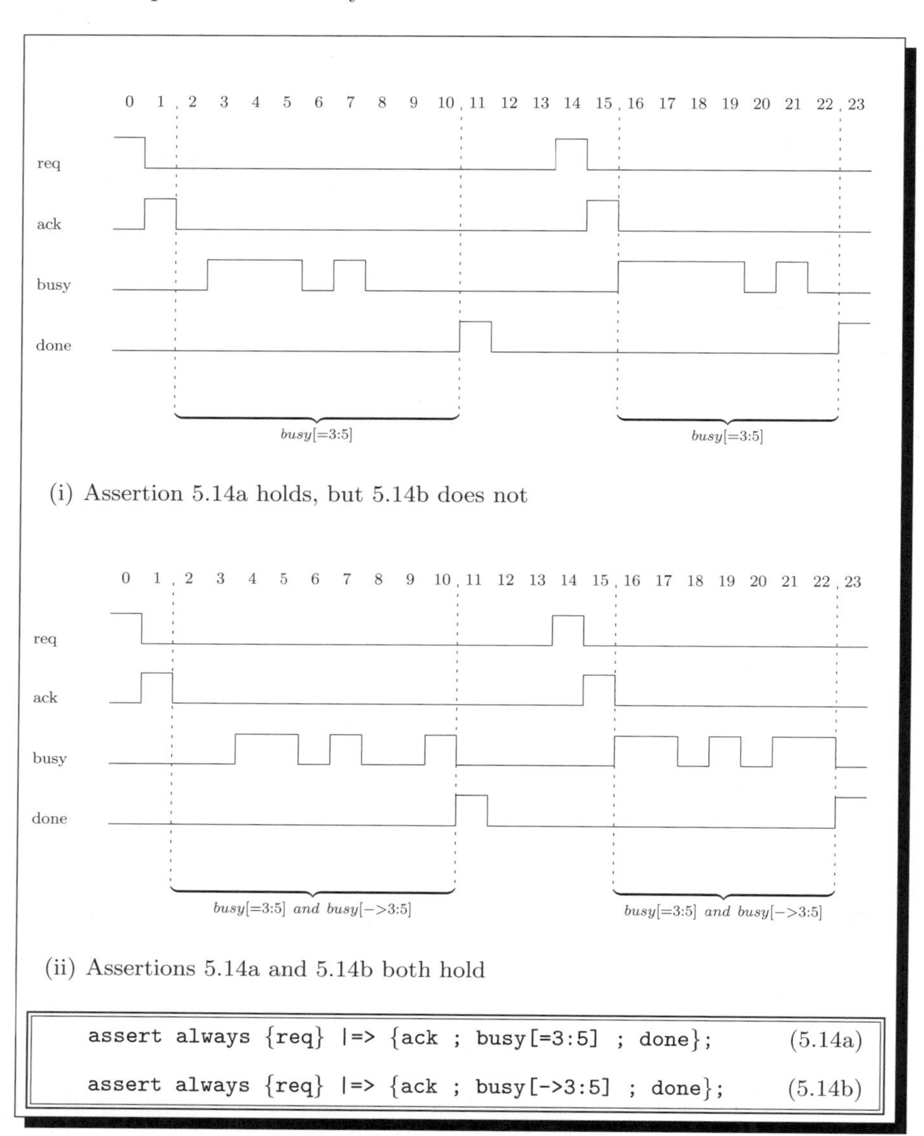


```
assert always {req} |=> {ack ; busy[=3:5] ; done};      (5.14a)
assert always {req} |=> {ack ; busy[->3:5] ; done};     (5.14b)
```

Fig. 5.14: The nonconsecutive and the goto repetition operators can take a range

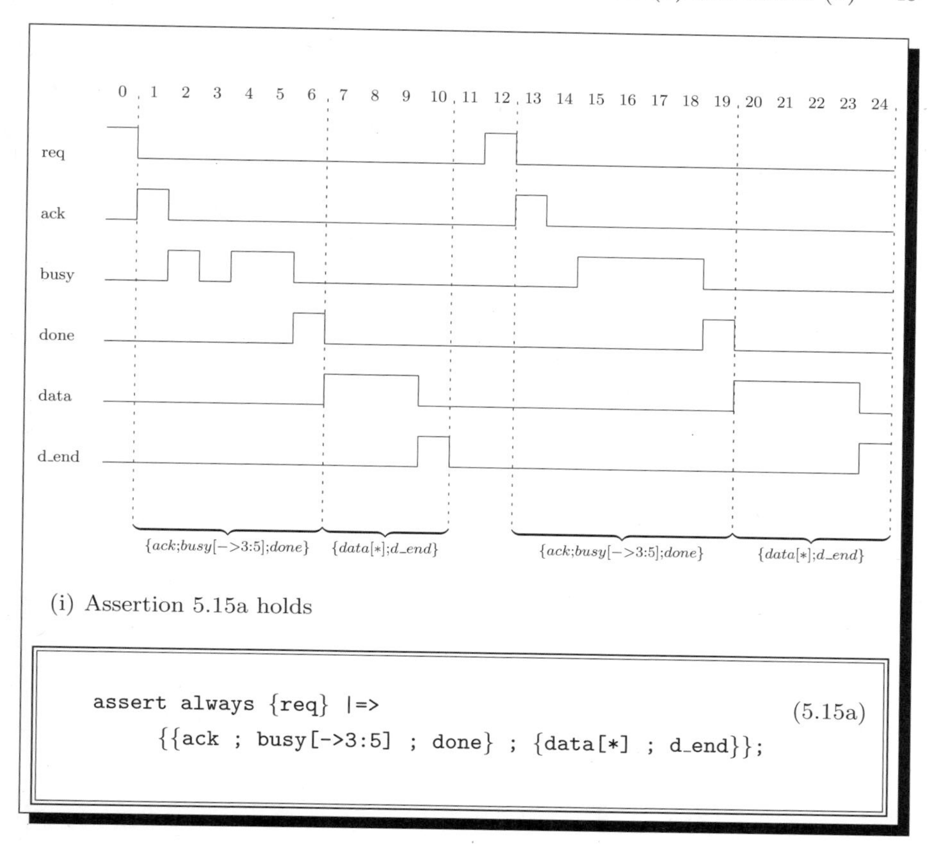

Fig. 5.15: Concatenation

If the data transfer should start the same cycle as signal `done` is asserted, then we can use the fusion operator as in Assertion 5.16a. Trace 5.16(i) is an example of a trace on which Assertion 5.16a holds. It is similar to Trace 5.15(i), except that the data transfer starts the same cycle as that in which `done` is asserted, rather than the cycle following the assertion of signal `done`.

Note that while it may be tempting to understand the concatenation operator (;) as a delay, the operator does not necessarily "eat" a cycle. For instance, in the case of the SERE {`a ; b[*] ; c`}, the `b[*]` may match zero, one, or more cycles, as illustrated in Trace 5.17(i). The first match shown in Trace 5.17(i) is only two cycles long, thus, the second concatenation operator (;) did not "eat" a delay. A better way to think of it is that concatenation gives you an ordered list of things to happen – some of them may consume one cycle, some more, and some no cycles at all.

Further note that fusion requires an overlap of at least one cycle. Thus, while the {`b[*]`} in {`b[*] ; c`} may match an empty sequence of cycles,

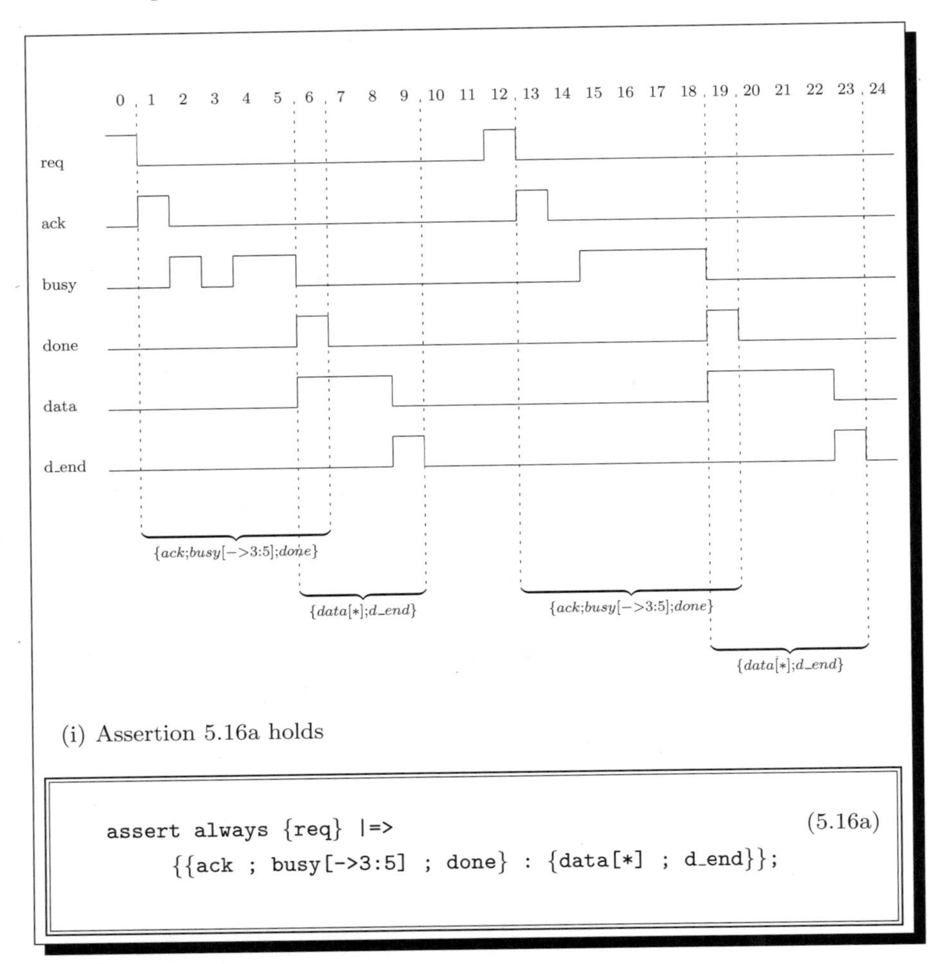

(i) Assertion 5.16a holds

Fig. 5.16: Fusion

replacing the concatenation operator with the fusion operator like this: `{b[*] : c}` results in a SERE with at least one assertion of `b` (otherwise there is nothing to overlap with the assertion of `c`). In both cases, the match may consist of a single cycle. The difference is that in the case of `{b[*] ; c}` the single cycle may not include an assertion of `b` (because there may be zero assertions of `b` preceding `c`), while in the case of `{b[*] : c}`, the single cycle must include an assertion of `b` (which overlaps with the assertion of `c`).

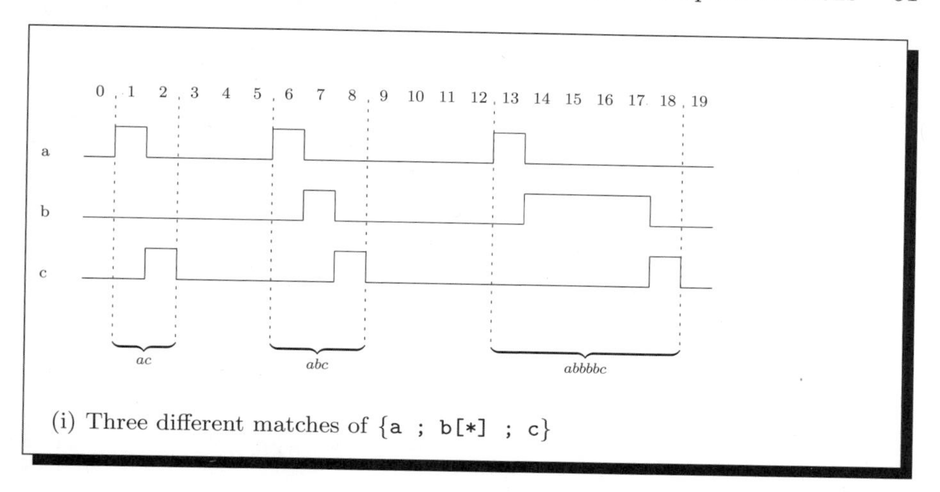

(i) Three different matches of {a ; b[*] ; c}

Fig. 5.17: The concatenation operator does not necessarily "eat" a delay

5.6 Compound SEREs

Compound SEREs are SEREs built from other SEREs through operations other than concatenation and fusion. The available operators allow you to "and" or "or" together two or more SEREs, as well as to express the situation where a match of one SERE occurs within a sequence of cycles matched by another.

Consider the case that we want to say that signal `read_complete` is asserted on the last data cycle of every read operation, where we have two kinds of read operations: a short read consists of an assertion of signal `short_rd` followed by eight not necessarily consecutive assertions of `data`, and a long read consists of an assertion of signal `long_rd` followed by 32 such assertions. We could code two separate properties, as shown in Properties 5.18a and 5.18b. Alternatively, we could use the SERE "or" operator (|) to "or" the two SEREs together, as in Property 5.18c.

```
always {short_rd ; data[->8]} |-> {read_complete}                (5.18a)

always {long_rd ; data[->32]} |-> {read_complete}                (5.18b)


always {{short_rd ; data[->8]} | {long_rd ; data[->32]}}          (5.18c)
   |-> {read_complete}
```

Fig. 5.18: The SERE "or" operator

The SERE "and" operator comes in two types: length-matching and non-length-matching. Both the length-matching (`&&`) and the non-length-matching (`&`) "and" operators match a sequence of cycles if starting at the current cycle, the left-hand side and the right-hand side are matched. The difference is that in addition, the length-matching "and" operator requires that the lengths of the sequences matched by both the left- and right-hand sides are the same, while the non-length-matching operator matches even if they are different. A length-matching "and" between SEREs `R` and `S` is illustrated in Trace 5.19(i), and a non-length-matching "and" between SEREs `R` and `S` is illustrated in Traces 5.20(i), 5.20(ii) and 5.20(iii).

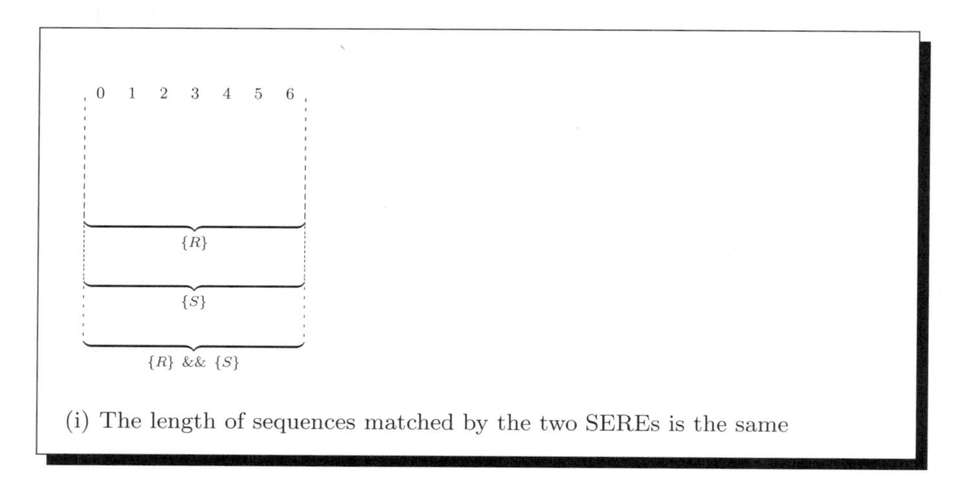

(i) The length of sequences matched by the two SEREs is the same

Fig. 5.19: Length-matching "and"

When a non-length-matching "and" is used on the left-hand side of a concatenation or a fusion, the current cycle of the right-hand side of the concatenation or fusion is determined by the longer of the two subsequences making up the non-length-matching "and". For instance, in the SERE `{{{a;b[*5];c} & {d[*];e}} ; f}`, the current cycle of `f` is the cycle after the cycle in which the longer of the two SEREs `{a;b[*5];c}` and `{d[*];e}` completes. This is illustrated in Traces 5.21(i) and 5.21(ii). Note that in each of the traces, signal `f` is asserted twice: once after the completion of `{a;b[*5];c}`, and once after the completion of `{d[*];e}`. However, in each trace, only the assertion of signal `f` that happens after the completion of the longer of the two SEREs participates in the "match" of SERE `{{{a;b[*5];c} & {d[*];e}} ; f}`.

NOTE: A length-matching "and" such as `{a;b;c} && {d}` is legal, but makes no sense (because there is no sequence which is both 3 cycles long to match `{a;b;c}` and 1 cycle long to match `{d}`). Many tools will probably issue a warning for such a SERE.

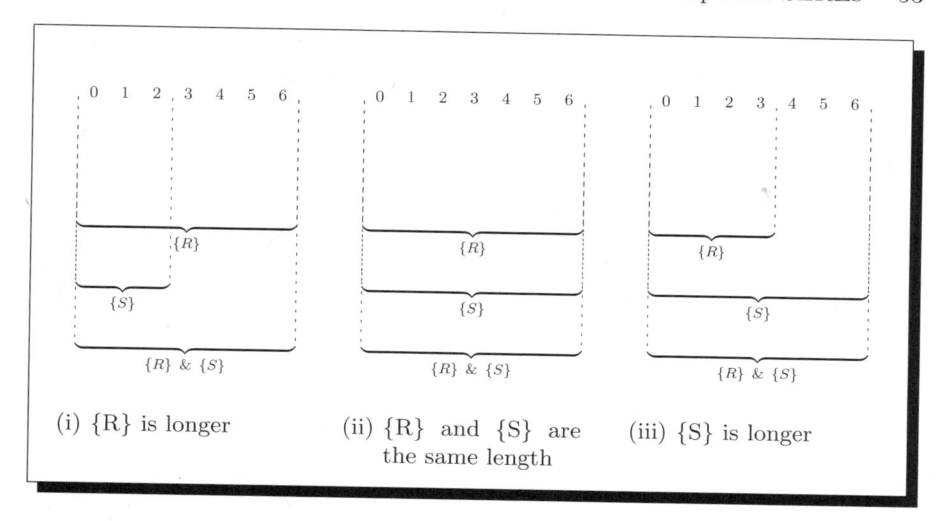

Fig. 5.20: Non-length-matching "and"

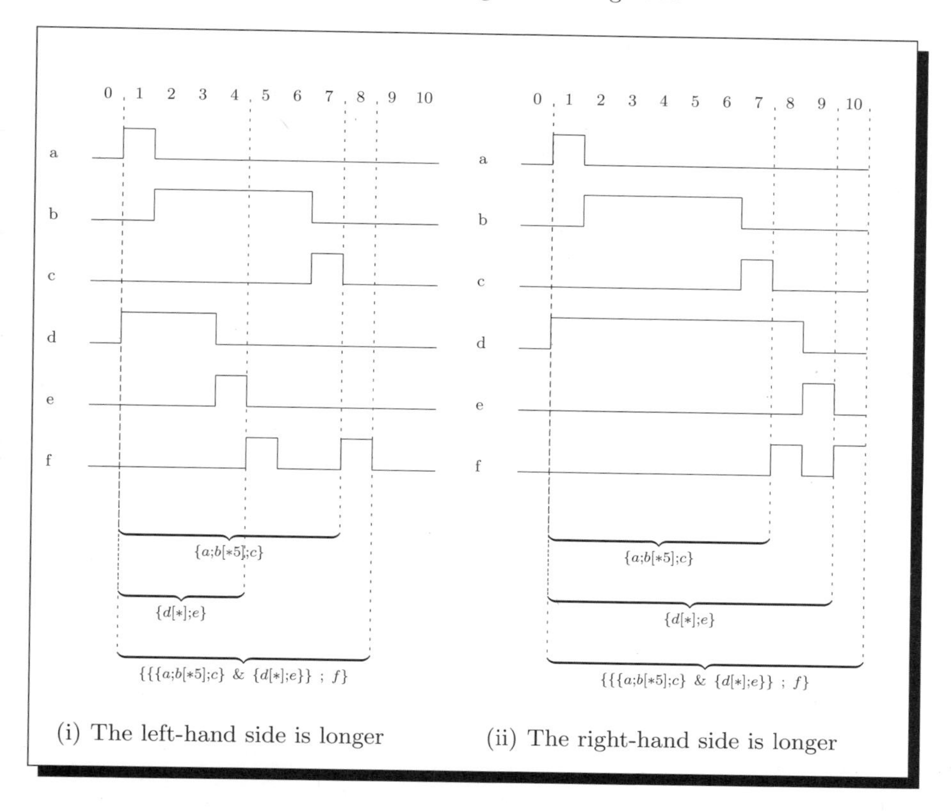

Fig. 5.21: Non-length-matching "and" on the left-hand side of a concatenation

As an example of the use of the length-matching "and", consider the case that a read request (assertion of signal **read_req**) that is granted (assertion of signal **gnt** within 5 cycles of the request) should be followed by a data transfer (assertion of signal **data_start** followed by some number of consecutive assertions of **data**, followed by an assertion of **data_end**), unless it is canceled. A cancel is an assertion of signal **cancel** any time between the assertion of **read_req** and the assertion of **gnt**, inclusive. We can express this using Assertion 5.22a. Assertion 5.22a holds on Trace 5.22(i): the first read request is followed by a data transfer, while the second read request is not followed by a data transfer, but is not required to be since it is canceled.

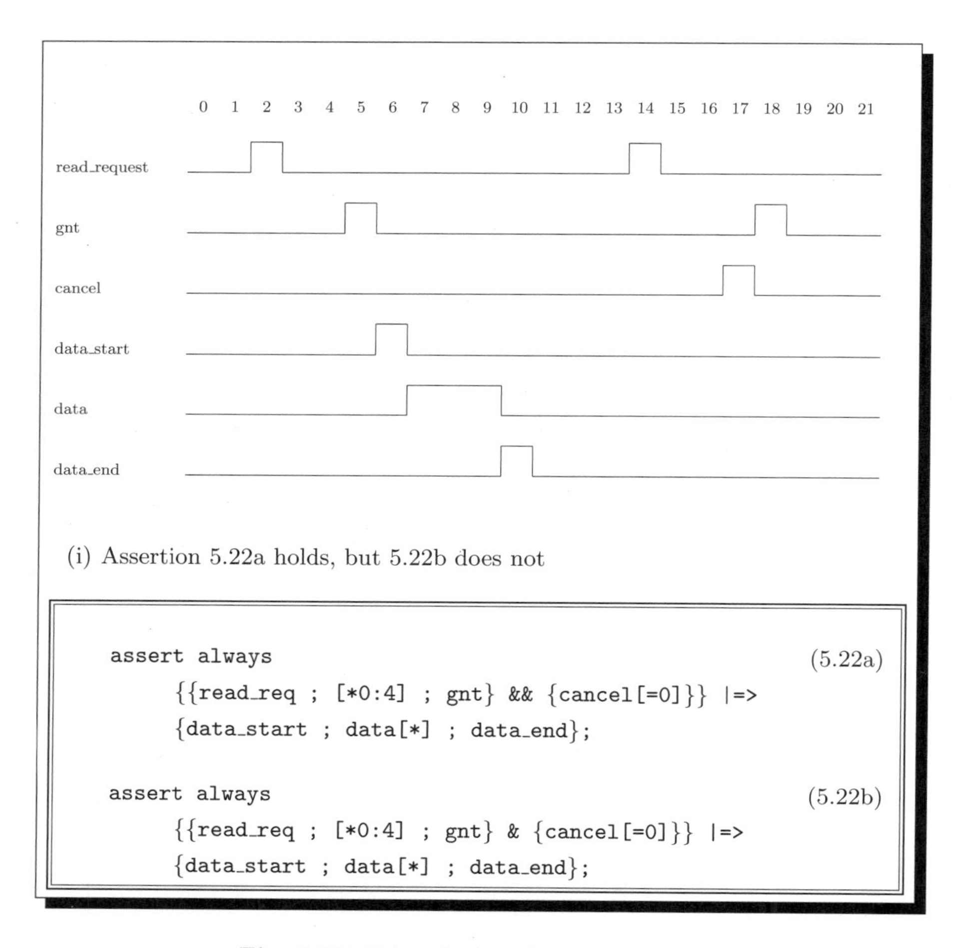

Fig. 5.22: Using the length-matching "and"

In the case of Assertion 5.22a, we needed the length-matching "and". To see why, consider what would have happened had we used the non-length-matching "and" as in Assertion 5.22b. In the case of Assertion 5.22b, the left-hand side SERE {{read_req ; [*0:4] ; gnt} & {cancel[=0]}} is matched not only by the sequence of cycles starting at cycle 2 and ending at cycle 5, but also by the sequences of cycles starting at cycle 2 and ending at any of the cycles 6 through 16. Thus, Assertion 5.22b requires that we see a data transfer starting not only at cycle 6, but also at cycles 7, 8, 9, 10 and so on. Obviously, this is not what we wanted.

As an example where we want the non-length-matching "and" rather than the length-matching "and", consider the case that signal global_print_req indicates that we should see a print on each of printers 1, 2, and 3 (completion of which is indicated by assertions of p1_done, p2_done, and p3_done, respectively), and that following completion of the last print job, we should see an assertion of signal print_success. We can express this as shown in Assertion 5.23a. Assertion 5.23a holds on Trace 5.23(i) because following the assertion of signal global_print_req, we see assertions of each of the signals p1_done, p2_done and p3_done, the last of which is followed by an assertion of print_success.

The non-length-matching "and" was needed for Assertion 5.23a. To see why, consider what would have happened if we had used the length-matching "and", as in Assertion 5.23b. Assertion 5.23b does not hold on Trace 5.23(i) because the use of the length-matching "and" means that p1_done[->], p2_done[->] and p3_done[->] must all have the same length, i.e., that p1_done, p2_done and p3_done must all be asserted at the same time.

The SERE within operator is useful if you want to describe a situation where one SERE occurs within the time frame of another. For instance, suppose that the normal behavior of the design is to complete a high priority request first, even if there is a pending low priority request that started before it. However, if signal no_nesting is asserted when the low priority request is issued, then this is prohibited. In other words, the situation shown in Trace 5.24(i) is not allowed. We can describe the prohibited situation as shown in Assertion 5.24a. Assertion 5.24a does not hold on Trace 5.24(i) because there is a match of {high_pri_begin ; high_pri_end[->]} that is entirely enclosed within the match of {(low_pri_begin && no_nesting) ; low_pri_end[->]}.

NOTE: If the left and right operands of a within operator are s and t respectively, then the within operator is simply a shorthand for a length-matching "and" between {[*] ; s ; [*]} and {t}. That is, s within t is equivalent to {[*] ; s ; [*]} && {t}.

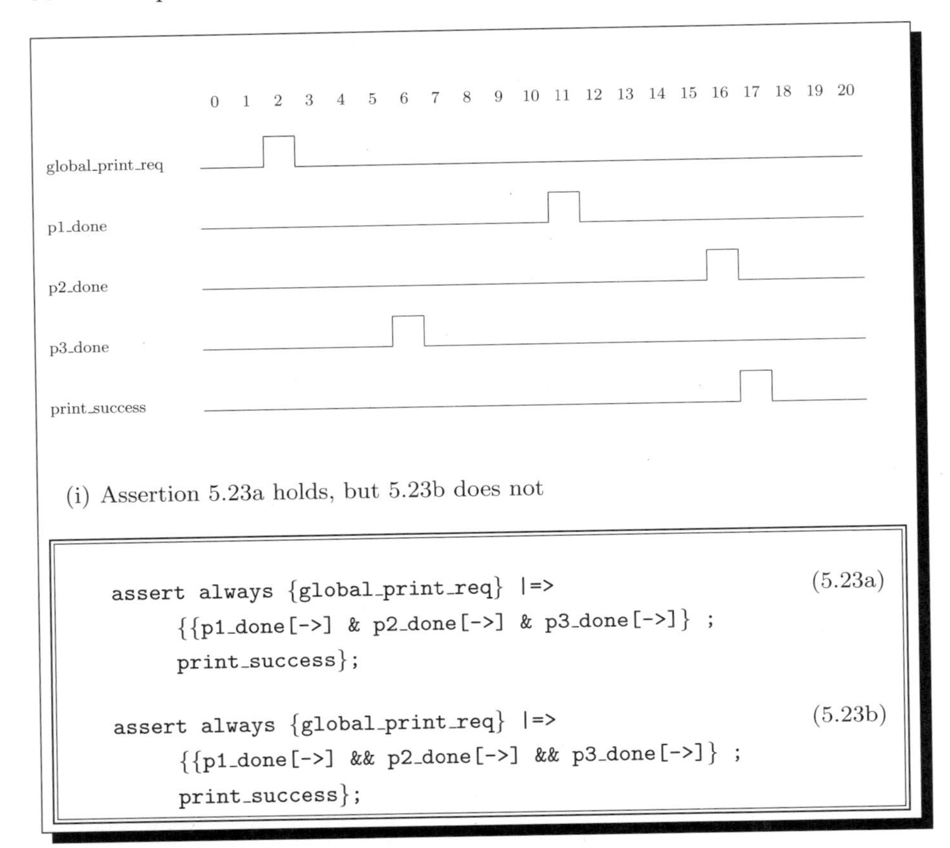

Fig. 5.23: Using the non-length-matching "and"

5.7 More about suffix implication

Up until now, we have seen the suffix implication operators (`|->` and `|=>`) used with SEREs on both the left- and the right-hand sides. While the left-hand side of a suffix implication operator must be a SERE, the right-hand side of a suffix implication operator can be any property. Thus, if we want to express the requirement that whenever we have a request that is acknowledged (assertion of `req` followed by an assertion of `ack`) we should see a grant (assertion of signal `gnt` within three cycles of the acknowledge), we can code as in Assertion 5.25a.

This is illustrated in Trace 5.25(i). In the trace, there are three occurrences of the SERE `{req;ack}`. The first is followed by an assertion of `ack` after two cycles, the second after three cycles, and the third after a single cycle. Thus, Assertion 5.25a holds on the trace.

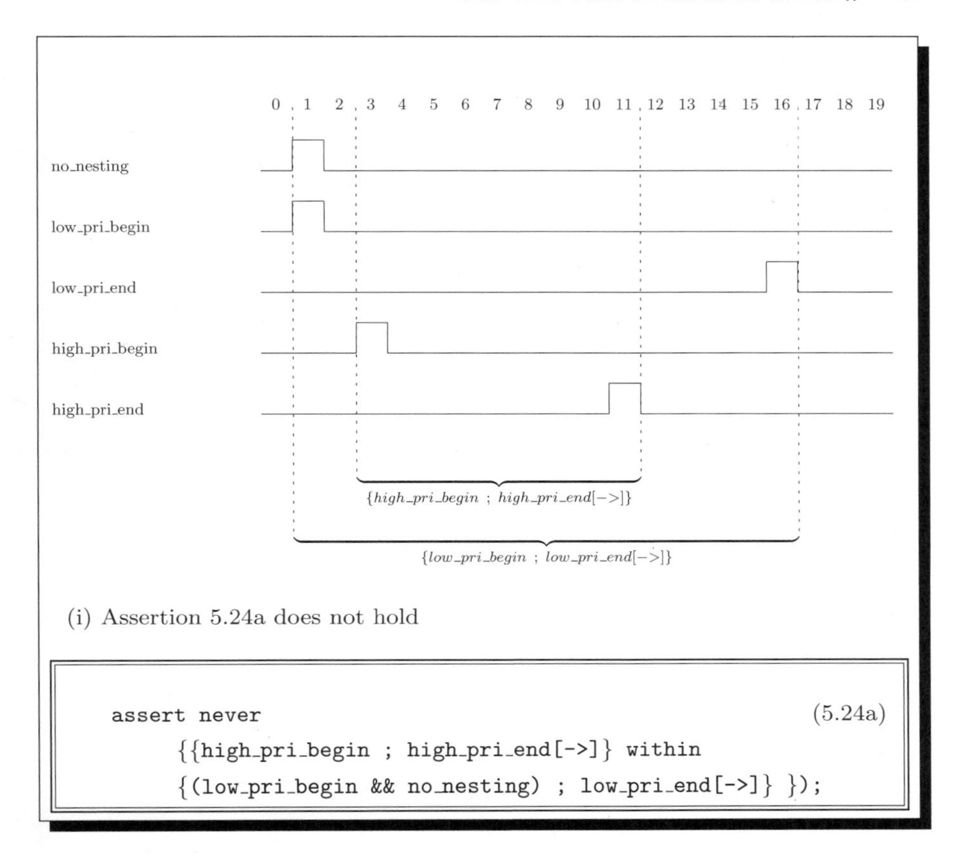

Fig. 5.24: The `within` operator

Of course, we could have written Assertion 5.25a entirely in SERE style, as shown in Assertion 5.25b, or entirely without SEREs, as in Assertion 5.25c. Assertions 5.25a, 5.25b and 5.25c are equivalent, thus the issue is purely one of style. Some people prefer using only one style, others find a mix of styles, depending on the particular property, to be easier to understand.

5.8 The built-in function `ended()`

The built-in function `ended()` takes a SERE as an argument and returns true in any cycle where that SERE has just ended.[1] For example, Trace 5.26(i) has

[1] In previous versions of PSL, the role of the built-in function `ended()` was played by *endpoints*. Endpoints are no longer a part of the language – use built-in function `ended()` instead.

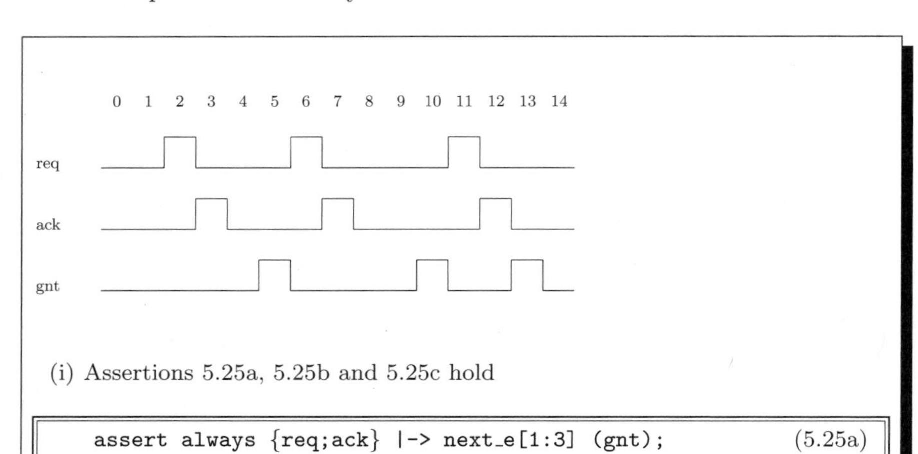

(i) Assertions 5.25a, 5.25b and 5.25c hold

```
assert always {req;ack} |-> next_e[1:3] (gnt);                (5.25a)
assert always {req;ack} |-> {[*1:3] ; gnt};                   (5.25b)
assert always (req -> next (ack -> next_e[1:3] (gnt)));       (5.25c)
```

Fig. 5.25: Three equivalent assertions

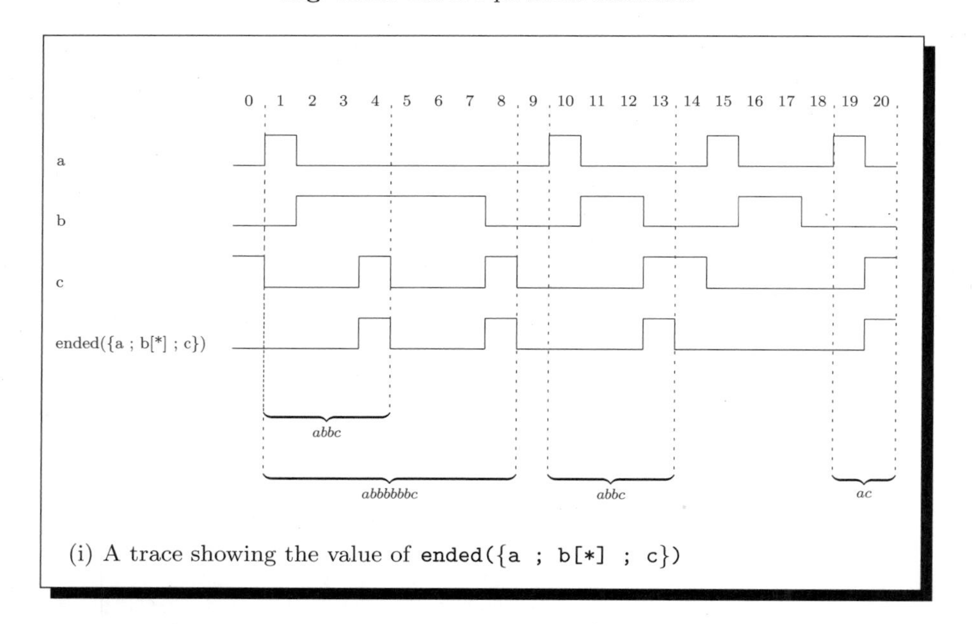

(i) A trace showing the value of `ended({a ; b[*] ; c})`

Fig. 5.26: `ended()`

been annotated with a waveform labeled `ended({a ; b[*] ; c})` to show the value of the call to `ended()` at each cycle.

How is `ended()` used? Consider that a complete data transfer consists of an assertion of signal `data_start` followed by some number of assertions of `data`, followed by an assertion of `data_end`. Consider further that we have a signal `data_transfer_complete` that should be asserted when a data transfer is completed, and that we would like to specify the correct behavior of this signal. A good start would be Assertion 5.27a. Assertion 5.27a ensures that `data_transfer_complete` is asserted at the conclusion of every data transfer. But what about the other direction? That is, what about ensuring that whenever `data_transfer_complete` is asserted, a complete data transfer has indeed concluded? Assertion 5.27a does not ensure this, and thus it holds on Trace 5.27(i), even though there are "extraneous" assertions of `data_transfer_complete` at cycles 6 and 15. We would like an assertion that holds on Trace 5.27(ii) but not on 5.27(i).

In order to check the second direction (that there are no extraneous assertions of `data_transfer_complete`) we need to switch the direction of the implication: we need to say that *if* `data_transfer_complete` is asserted, *then* we have just seen the conclusion of a complete data transfer. We could try Assertion 5.27b. However, that doesn't work. Assertion 5.27b requires that the SERE {`data_start ; data[*] ; data_end`} *start* the same cycle as `data_transfer_complete`, while we want it to *end* that cycle.

This is where the built-in function `ended()` comes in. Using `ended()`, we can code what we want as shown in Assertion 5.27c, which states that whenever we see an assertion of `data_transfer_complete`, we must have just seen the end of SERE {`data_start ; data[*] ; data_end`}, and it does not hold on Trace 5.27(i).

Note that `ended()` returns a Boolean value. Since the left-hand side of the suffix implication in Assertion 5.27c consists of a single Boolean expression, we could have coded Assertion 5.27c equivalently using logical implication as shown in Assertion 5.27d.

5.9 Overlapping matches of a SERE

Up until now, we have been concentrating on explaining the meaning of each SERE operator, using relatively simple examples. It is now time to introduce some more complex examples, in order to emphasize that like LTL style properties, interpreting a SERE style property on a trace can involve examining overlapping sets of cycles.

Consider Property 5.28a on Traces 5.28(i) and 5.28(ii). There are multiple matches of {`a ; b[*] ; c`} on Trace 5.28(i), one match ending at cycle 2, one ending at cycle 3, and one ending at cycle 4. Property 5.28a does not hold on Trace 5.28(i), because the matches ending at cycles 2 and 3 do not have associated assertions of signal `d` at cycles 3 and 4. A suffix implication

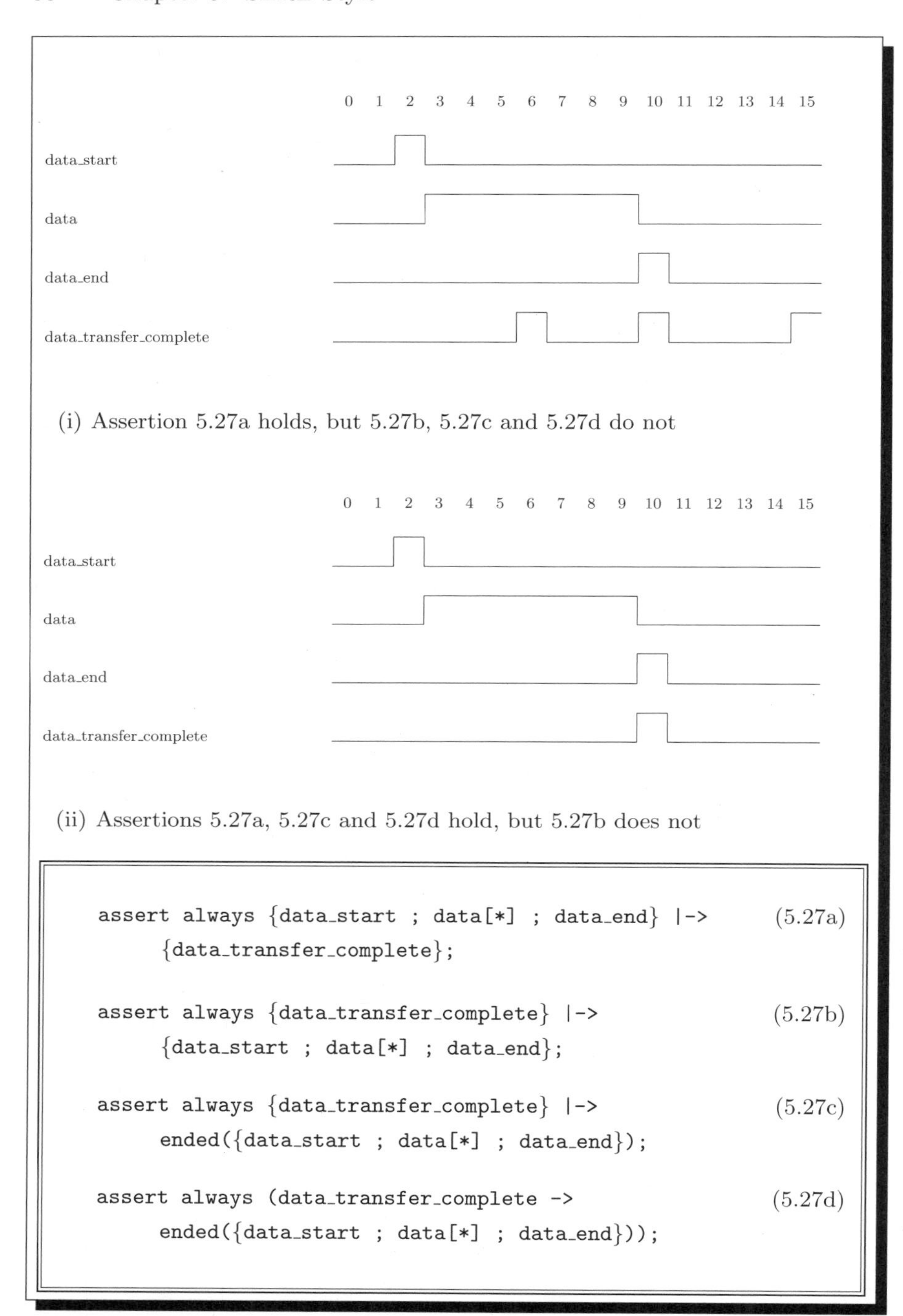

Fig. 5.27: Using `ended()`

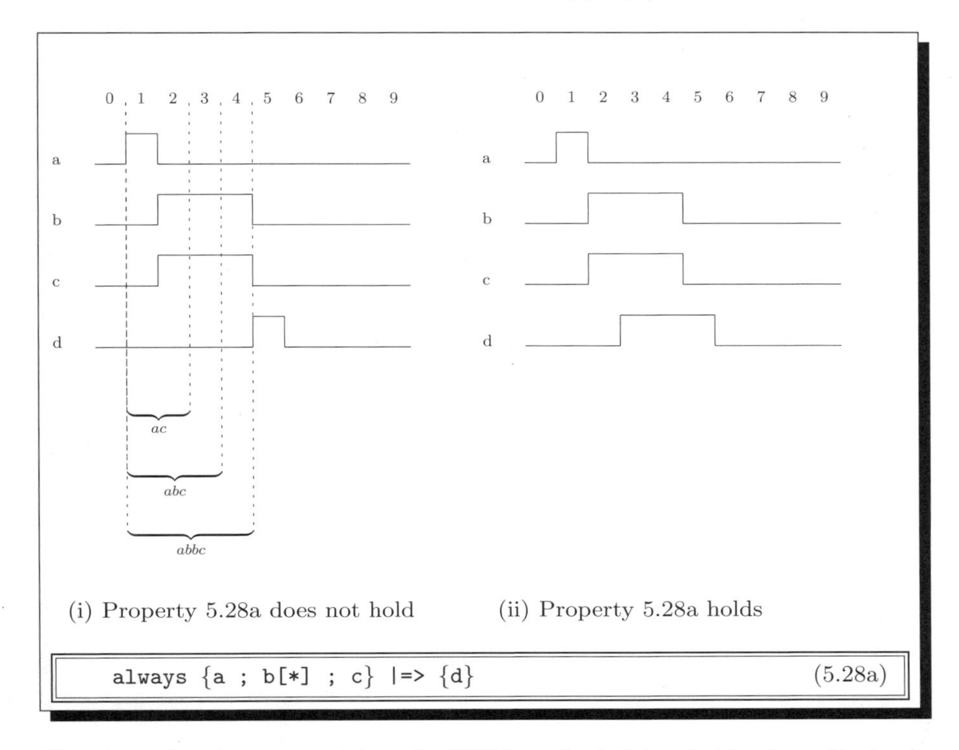

(i) Property 5.28a does not hold

(ii) Property 5.28a holds

```
always {a ; b[*] ; c} |=> {d}                    (5.28a)
```

Fig. 5.28: Overlapping matches of a SERE on the left-hand side of a suffix implication

requires that for *every* match of the left-hand side, the right-hand side holds. Property 5.28a does hold on Trace 5.28(ii), because signal `d` is asserted after each match of {`a ; b[*] ; c`}.

Note that although Property 5.28a contains an `always` operator, the `always` operator is not the source of the overlapping matches in this case. Rather, for a single current cycle (cycle 1), there are three separate matches of the SERE {`a ; b[*] ; c`}, each ending at a different cycle.

The overlapping can happen on the right-hand side of a suffix implication as well. Consider for example Property 5.29a on Trace 5.29(i). There are three matches of {`d ; e[*] ; f`}, ending at cycles 3, 4, and 5. One match would have been enough to satisfy Property 5.29a, because as long as there exists one match, {`d ; e[*] ; f`} holds. The fact that there are in fact three matches does not hurt. Thus, Property 5.29a holds on Trace 5.29(i).

We have just explained that we require *at least one* match of the right-hand side for *every* match of the left-hand side of a suffix implication, and it might seem that in so doing, we have introduced something new. Actually, the rule could have been deduced from the parallel we have previously drawn between a suffix implication and an if-then statement. SERE {`a ; b[*] ; c`} can be

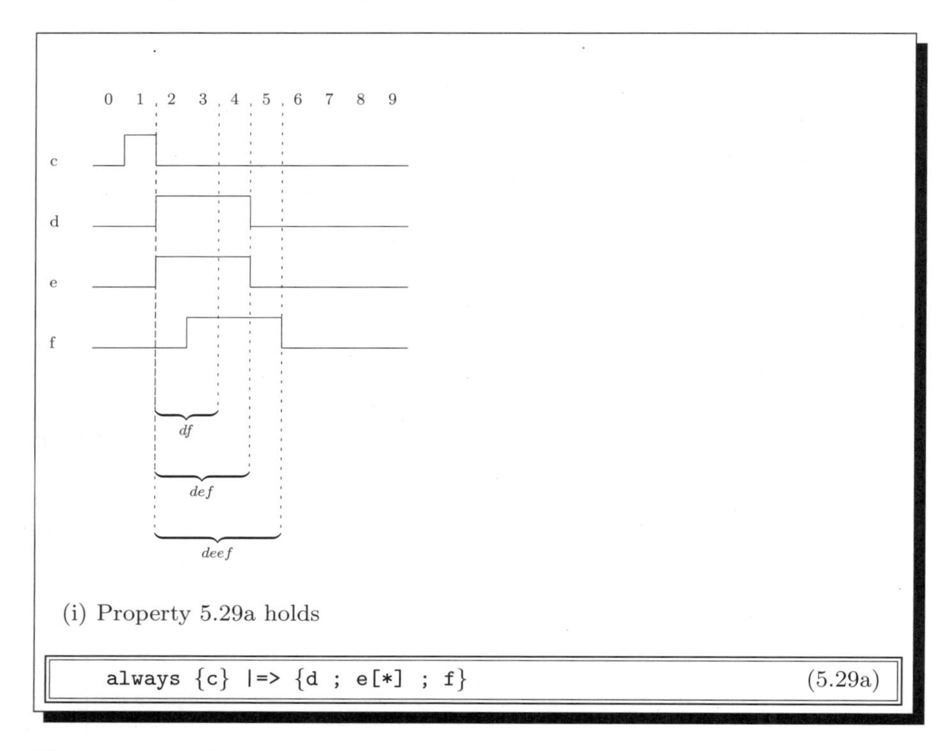

Fig. 5.29: Overlapping matches of a SERE on the right-hand side of a suffix implication

understood as an infinite "or" between SEREs {a;c}, {a;b;c}, {a;b;b;c}, and so on. So Property 5.28a can be understood as "if {a;c} or {a;b;c} or {a;b;b;c} or ..., then d", which clearly should hold only if we see a d every time the if-part is matched. And Property 5.29a can be understood as "if c, then {d;f} or {d;e;f} or {d;e;e;f} or ...", which of course should hold as long as we have seen at least a single match of the then-part.

5.10 How not to use SEREs

A SERE is a property on its own. However, a stand-alone SERE is not a very interesting property. For example, if we assert the property formed from a single SERE, as in Assertion 5.30a, we are stating that we require that (req_out && !ack) be asserted on the first cycle, then busy && !ack be asserted for zero or more cycles, and finally that ack be asserted (if busy && !ack does not hold forever). Similarly to the assertion assert a;, this is not an assertion that would be useful in many designs. Furthermore, adding an always operator to Assertion 5.30a does not make the assertion any more

```
assert {(req_out && !ack) ; (busy && !ack)[*] ; ack};       (5.30a)

assert always                                               (5.30b)
    {(req_out && !ack) ; (busy && !ack)[*] ; ack};
```

Fig. 5.30: Don't use SEREs like this

useful. Assertion 5.30b states that we expect the SERE {(req_out && !ack) ; (busy && !ack)[*] ; ack} to hold at every cycle. Thus, at every cycle, we expect (req_out && !ack) to be asserted, followed by zero or more assertions of busy && !ack followed finally by an ack in the case that busy && !ack does not hold forever. In particular, this means that req_out must be asserted every cycle, and ack deasserted every cycle. As with Assertion 5.30a, this is not an assertion that is particularly useful in many designs. Thus, the typical use of a SERE will be either as the left- or right-hand side of a suffix implication or as the operand of a never operator.

6

Clocks

Up until now, we have examined cycle-based traces – we have not explicitly mentioned the clock. The clock operator allows you to describe traces that are based on "ticks" of one or more clock signals. For example, consider the requirement "two consecutive requests (assertions of signal `req`) are not allowed" in a design clocked on the rising edge of signal `clk` that behaves as shown in Trace 6.1(i). There are three "ticks" of the clock signal `clk` in Trace 6.1(i), each of them three cycles wide.

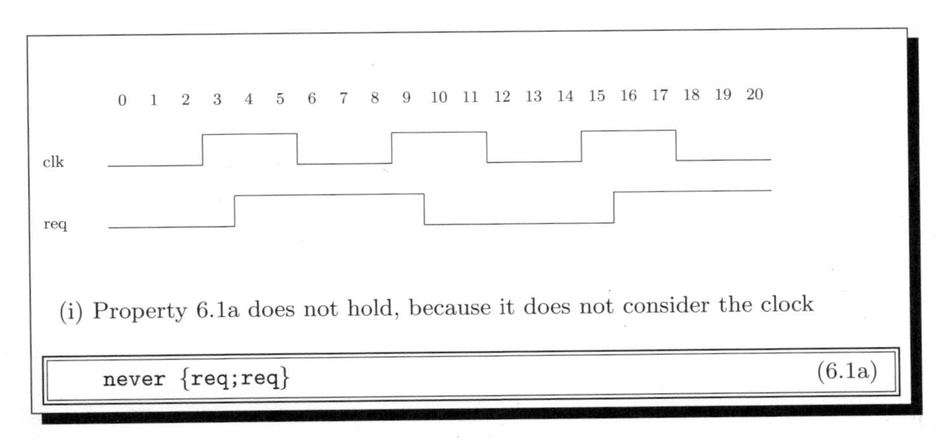

Fig. 6.1: A trace clocked on the rising edge of signal `clk`

NOTE: The term *cycle* in PSL is defined as one iteration of the evaluation process, during which the state of the design is recomputed and may change. In other words, a cycle is the smallest granularity of time as seen by the trace in question. In a cycle-based environment, then, a PSL cycle corresponds to a clock cycle. In an event-based environment, a PSL cycle is different than a clock cycle. We will use the term *cycle* when we mean a PSL cycle, and *clock cycle* when we mean a clock cycle.

Getting back to our example, if we code an unclocked property expressing our specification, as in Property 6.1a, we find that it does not hold on Trace 6.1(i) (because `req` is asserted at the consecutive cycles 4 and 5, 5 and 6, 6 and 7, etc.). Despite the fact that Property 6.1a does not hold on Trace 6.1(i), clearly Trace 6.1(i) adheres to the requirement that two consecutive requests are not allowed.

6.1 Edge-triggered designs

What is missing from Property 6.1a, repeated in Property 6.2a, is the notion of the clock. Let's clock it using the clock operator `@`, as shown in Property 6.2b. The call to the built-in function `rose(clk)` returns true if `clk` is asserted and was deasserted the previous (PSL) cycle. The `@rose(clk)` in Property 6.2b has the effect of filtering out all but the cycles on which `rose(clk)` holds. The clocked property holds when the unclocked property holds on the cycles left after the filtering. This is illustrated in Trace 6.2(i). The clock expression `rose(clk)` holds at cycles 3, 9, and 15, shaded in Trace 6.2(i). Thus, Property 6.2b holds on Trace 6.2(i) if the unclocked version, Property 6.2a, holds on the trace composed of only these cycles. It does, and thus Property 6.2b holds on Trace 6.2(i).

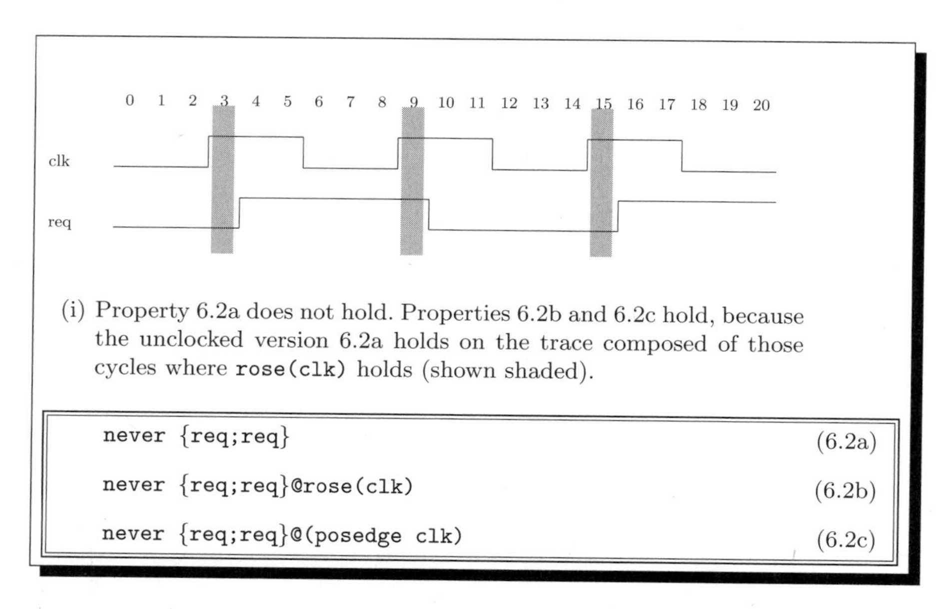

(i) Property 6.2a does not hold. Properties 6.2b and 6.2c hold, because the unclocked version 6.2a holds on the trace composed of those cycles where `rose(clk)` holds (shown shaded).

`never {req;req}` (6.2a)

`never {req;req}@rose(clk)` (6.2b)

`never {req;req}@(posedge clk)` (6.2c)

Fig. 6.2: Trace 6.1(i) shaded to show the cycles where `rose(clk)` holds

As we have seen, clocking with `rose(clk)` is the way to clock properties in a design with flip-flops triggered by the rising edge of `clk`. As we would expect, a design with flip-flops triggered on the falling edge of `clk` should be clocked with `fell(clk)`. In the Verilog flavor, we can equivalently clock using `@(posedge clk)` and `@(negedge clk)`.[1] In the explanation of why Property 6.2b holds on Trace 6.2(i), we said that the clock expression `rose(clk)` holds at cycles 3, 9 and 15, and we have just claimed that Property 6.2b can be equivalently expressed in the Verilog flavor as shown in Property 6.2c. This might seem a little strange. An edge is something that happens between cycles – how can it occur *on* a cycle? The answer is that in PSL, clocking a property with `@(posedge clk)` can be understood as "look at those cycles that it makes sense to look at in a design clocked on the positive edge of signal `clk`", and of course similarly for `@(negedge clk)`. No matter the width of the clock, the cycles immediately after an edge are guaranteed to hold a steady-state value, thus these are the cycles used in determining whether or not a property clocked with `@(posedge clk)` or `@(negedge clk)` holds.

Let's consider another example. Suppose that we want to write a property that states that signal `b` is a latched version of signal `a`. We can code Property 6.3a and we expect that it should hold on Trace 6.3(i), and indeed, it does.[2] But if you are used to looking at timing diagrams for a living, you probably looked at (PSL) cycles 2 and 4 to check the property for the first clock cycle, at cycles 8 and 10 for the second, and so on, while above, we explained that PSL will be looking at the cycles immediately following a rising edge, that is, at cycles 3, 9, 15, and 21, shown shaded in Trace 6.3(ii).

Why do we recommend looking at cycles 3, 9, 15, and 21, when intuitively, it makes more sense to look at the pairs (2,4), (8,10), etc? The answer is that the two ways of looking at things are equivalent, as long as the signals being examined are glitch-free – that is, as long as they change only once during a clock cycle. Thus, cycles 0 through 3 are equivalent for signal `a` because it has a steady value from cycles 0 through cycles 3. Cycles 4 through 9 are equivalent, as are cycles 10 through 15, and so on.

The bottom line is that PSL defines that `@(posedge clk)` and `@(negedge clk)` filter out all cycles except those immediately following a clock edge, but as long as your design is glitch-free you can, when interpreting a trace, look at the equivalent cycles immediately preceding the clock edge, or at pairs of cycles, one of which precedes the edge and one of which follows it, like the pair (8,10) in Trace 6.3(i).

---

[1] To be exact, we must specify that the equivalence holds only if signal `clk` takes on only the values 0 and 1 (i.e., not X or Z), and no signal in the property being clocked changes at the same time as `clk` (i.e., there are no race conditions).

[2] Our property checks that when `a` is asserted, `b` will be asserted the following cycle. To completely describe the design, we should also state the similar property that when `a` is deasserted, `b` is deasserted the following cycle. In the interest of brevity, we omit the second property throughout this chapter.

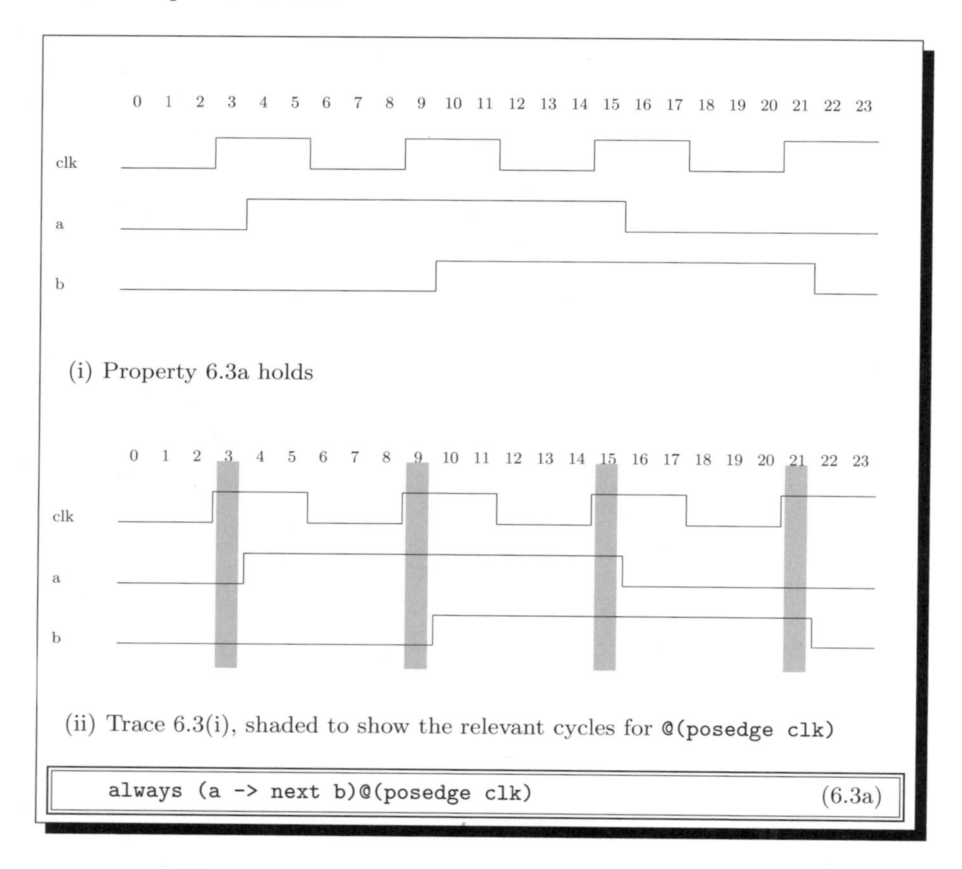

Fig. 6.3: Stating that signal `b` is a latched version of `a`

6.2 Level-sensitive designs

How should properties be clocked for a level-sensitive design? Let's start with how they should *not* be clocked. The property `always a@clk` is a valid PSL property. However, don't confuse the PSL `@clk` construct with an event trigger of Verilog. The Verilog code shown in Figure 6.4 will assign the value of `b` to `a` whenever a change in the value of signal `clk` is detected. In PSL, the `@clk`

```
always @(clk)
  a <= b;
```

Fig. 6.4: Some Verilog code

construct does not indicate a filtering on the basis of a change in the value of `clk`; rather, it filters out all cycles except those on which signal `clk` is high.

A design using transparent latches sensitive to the high value of signal `clk` should *not* be clocked using `@clk`. This is because filtering out all but the cycles where `clk` holds is not what we want in a level-sensitive design. To see this, consider Property 6.5a in a design using transparent latches sensitive to the high value of the clock, as in Trace 6.5(i).

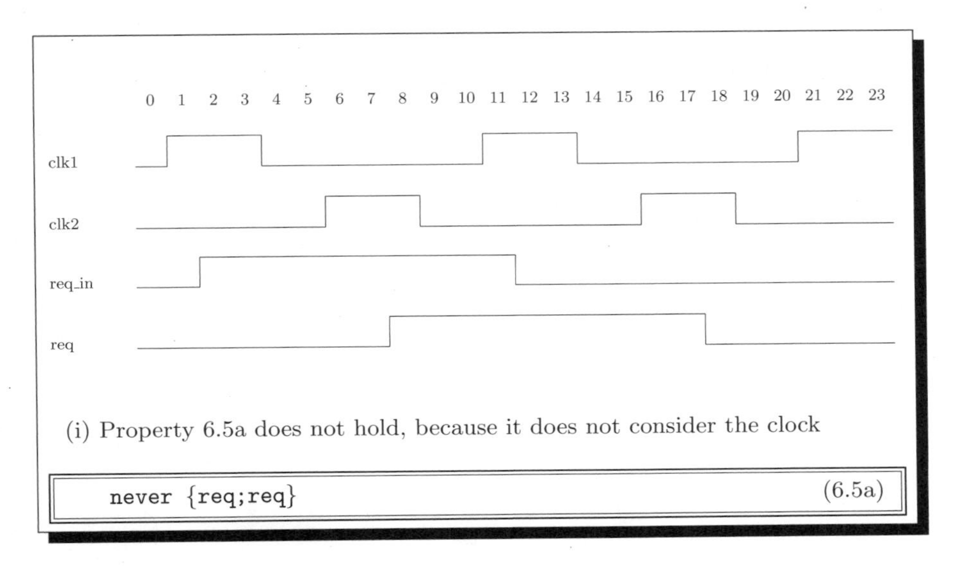

Fig. 6.5: A trace from a design using level-sensitive latches

As in most designs using level-sensitive latches, the design in Trace 6.5(i) has two clocks – `clk1` and `clk2`. The signal `req_in` is sensitive to `clk1`, while the signal `req` samples `req_in` at `clk2`. Although the design is level-sensitive, the trace looks similar to one that would result from an edge-triggered design. This is due to the relation between the clocks together with the fact that `req`, sensitive to `clk2`, is driven by `req_in`, sensitive to `clk1`.

Signal `req` is asserted for only one clock cycle in Trace 6.5(i), thus we would like a clocked version of Property 6.5a to hold on the trace. However, clocking Property 6.5a with `@clk1`, as in Property 6.6a, will have the effect of filtering out all but those cycles in which `clk1` holds, as shown in Trace 6.6(i). Property 6.6a does not hold on Trace 6.6(i) because signal `req` holds on consecutive shaded cycles (11,12) and (12,13). Thus, clocking with `@clk1` is not what we want. Similarly, we do not want to clock with `@clk2`, as in Property 6.6b, as illustrated in Trace 6.6(ii). Property 6.6b does not hold on Trace 6.6(ii) because signal `req` holds on consecutive shaded cycles (8,16) and (16,17), which is again not what we want.

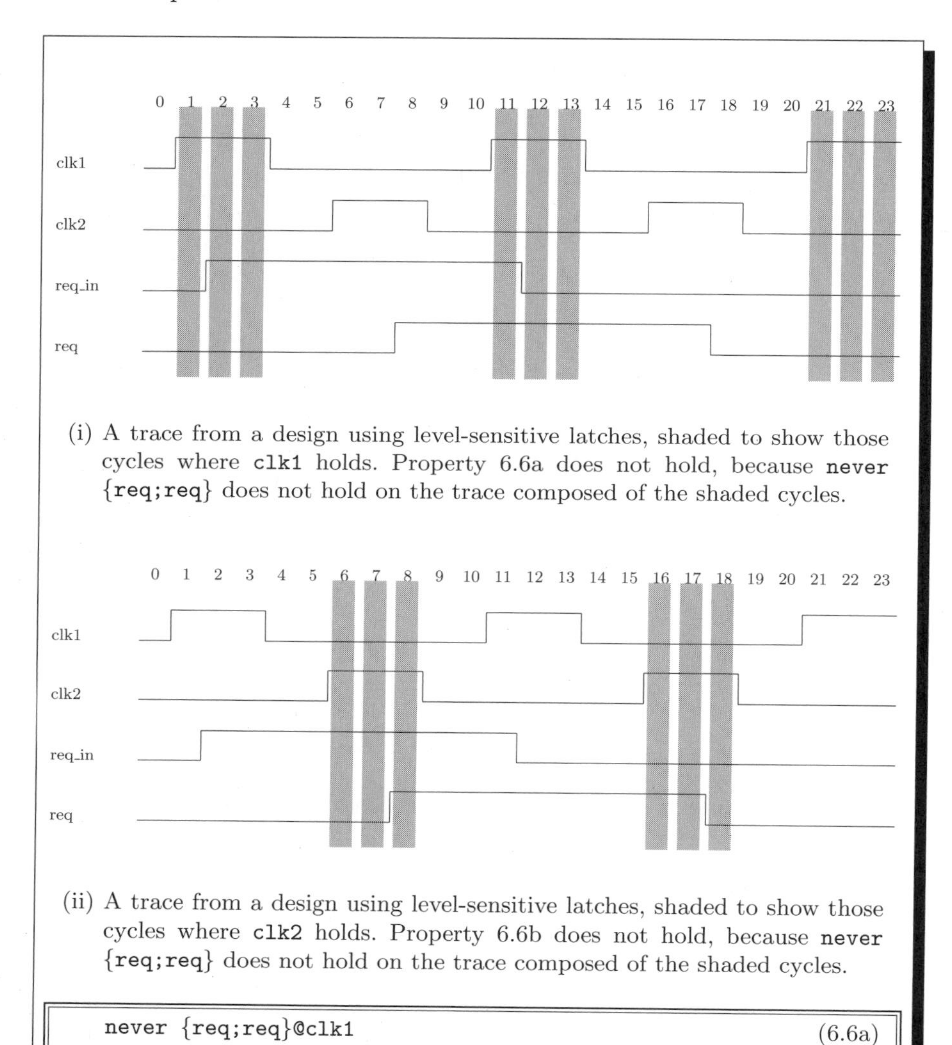

(i) A trace from a design using level-sensitive latches, shaded to show those cycles where `clk1` holds. Property 6.6a does not hold, because `never {req;req}` does not hold on the trace composed of the shaded cycles.

(ii) A trace from a design using level-sensitive latches, shaded to show those cycles where `clk2` holds. Property 6.6b does not hold, because `never {req;req}` does not hold on the trace composed of the shaded cycles.

```
never {req;req}@clk1                                        (6.6a)

never {req;req}@clk2                                        (6.6b)
```

Fig. 6.6: Level-sensitive latches – what not to do

We have seen that using `@clk1` or `@clk2` is not what we want. Recall that a level-sensitive design is usually two-phased, as shown in Traces 6.6(i) and 6.6(ii). In such a design, it is sufficient to look at a single (PSL) cycle of each clock cycle. For instance, we could have chosen the first (PSL) cycle of `clk2`, because at that cycle signals such as `req` that are sensitive to the high value of signal `clk2` are guaranteed to be at a steady state. Thus, in our level-sensitive design, we could have chosen to code as in Property 6.7a. This would have selected the cycles as shown in Trace 6.7(i).

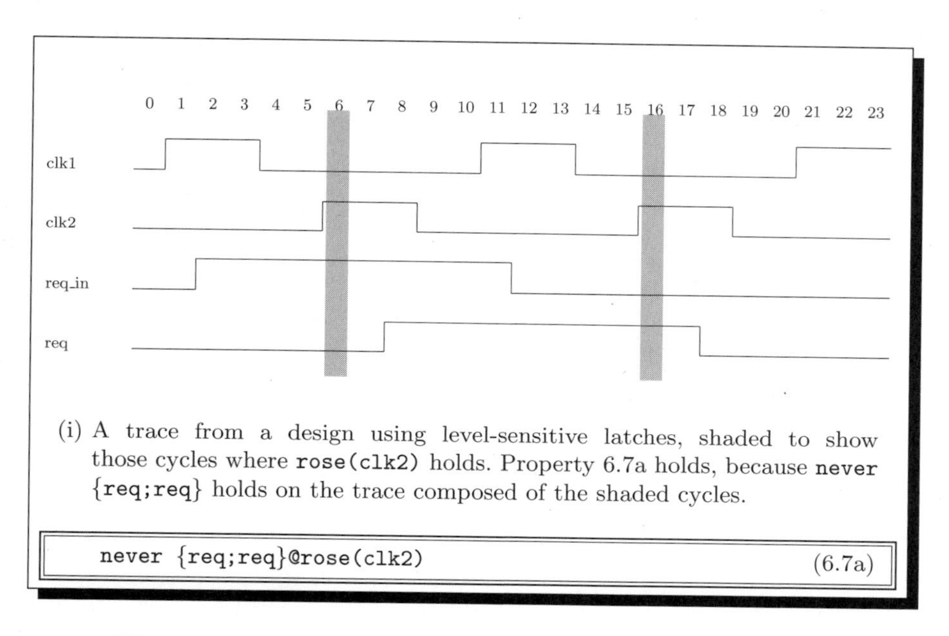

(i) A trace from a design using level-sensitive latches, shaded to show those cycles where `rose(clk2)` holds. Property 6.7a holds, because `never {req;req}` holds on the trace composed of the shaded cycles.

```
never {req;req}@rose(clk2)                                    (6.7a)
```

Fig. 6.7: Correct clocking of a design using level-sensitive latches

As we did for edge-triggered designs above, let's consider another example. Suppose that we want to write a property that states that signal `b` is a latched version of signal `a`. Since we have a two-phased design, we have to clarify exactly what we mean. If by "latched version" we mean that `a` is clocked on `clk1` and `b` is clocked on `clk2` (as shown in Figure 6.8(i)), we will need to code differently than if we mean that both are clocked on the same clock (as shown in Figure 6.8(ii)). Let's assume that we mean the latter, and defer the discussion of multiply-clocked properties to Chapter 14.

If we mean the situation shown in Figure 6.8(ii), where we do not involve signals latched on `clk2`, the situation is similar to that of a singly-clocked design, clocked on `clk1`. Thus we can code as shown in Property 6.9a, and we expect that Property 6.9a should hold on Trace 6.9(i) – and indeed, it does.

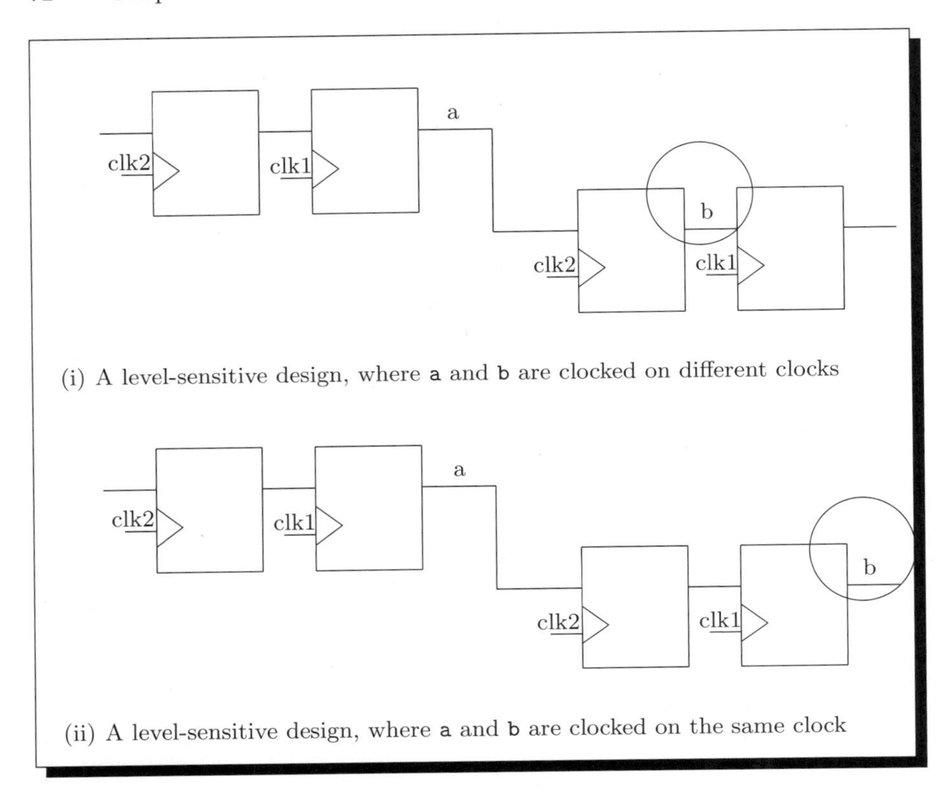

Fig. 6.8: Two multiply-clocked designs. The difference is the location of signal `b`.

We have seen that in a level-sensitive design, it is usually the case that you will clock with `@rose(clk)` or `@fell(clk)`, and previously we have discussed the fact that the same is true for edge-triggered designs. Furthermore, we have seen that clocking with `@(posedge clk)` is almost (see footnote 1 on p. 67) equivalent to clocking with `@rose(clk)`, and that clocking with `@(negedge clk)` is almost equivalent to clocking with `@fell(clk)`. Does it follow that we can clock a level-sensitive design with `@(posedge clk)` or `@(negedge clk)`? The answer is yes, although doing so is not recommended because it can be confusing for someone trying to understand the specification to see the use of `posedge` or `negedge` in a design that is not edge-triggered.

What, then, is the difference between a level-sensitive and an edge-triggered design? The difference is in the implementation, not in the functionality that we are trying to achieve. In both cases, there is the notion of a clock cycle, and in both cases, we need to look at a single (PSL) cycle of each clock cycle in order to understand what is going on. In PSL, you use the clock operator to specify which (PSL) cycles should be looked at, not

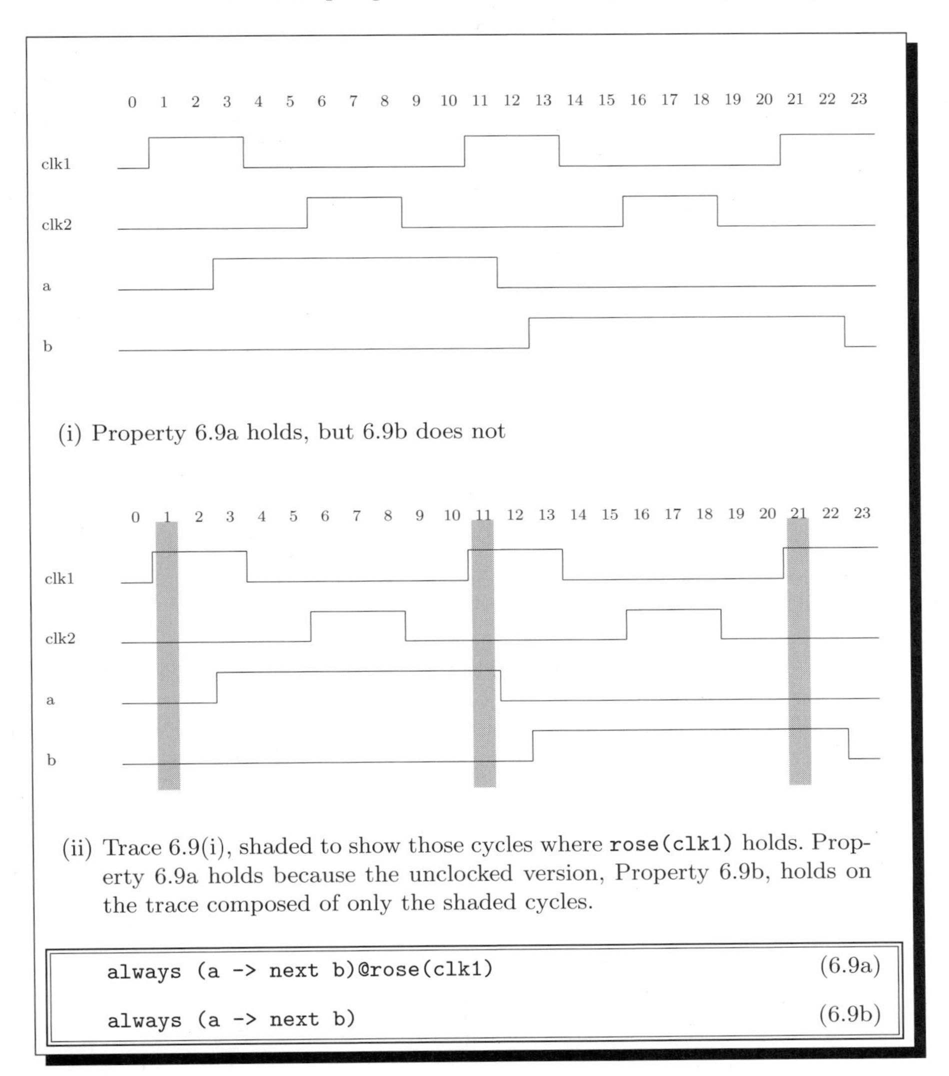

(i) Property 6.9a holds, but 6.9b does not

(ii) Trace 6.9(i), shaded to show those cycles where `rose(clk1)` holds. Property 6.9a holds because the unclocked version, Property 6.9b, holds on the trace composed of only the shaded cycles.

```
always (a -> next b)@rose(clk1)        (6.9a)

always (a -> next b)                   (6.9b)
```

Fig. 6.9: Signals `a` and `b` are clocked on the same clock

whether or not your design was implemented using edge-triggered flip-flops or level-sensitive latches.

6.3 Sampling semantics and the granularity of a cycle

PSL does not dictate how time ticks for an unclocked property. Rather, the smallest granularity of time is that seen by the verification tool. This separation of powers is intentional and extremely important: it means that the questions "How does the design behave in the verification tool when I write `always (prop)@(posedge clk)`?" and "What is the value of the property `always (prop)@(posedge clk)` in my design?" have very simple answers. The answer to the former question is that the design behaves however it would have behaved without PSL. In particular, a user of PSL does not have to worry about a property influencing the behavior of the design – PSL defines that such influence does not happen.

The answer to the latter question is: take the trace produced by the design (this trace is the same as the trace that would have been produced without PSL!), then evaluate the PSL property on that trace. In particular, a trace that has been saved during a verification run can be postprocessed to evaluate a set of PSL properties, and the result will be the same as the result that would have been calculated had the PSL properties been checked at runtime.

Note that it follows from PSL's policy on the granularity of time that two implementations may lead to different results for an unclocked property. In this way, it is possible to code a property that identifies a glitch during simulation (see Section 6.5), or one that is sensitive to the order in which signals are evaluated in an event-driven simulation.

While the smallest granularity of time as seen by PSL changes from tool to tool, the granularity of time as seen by a particular property can be controlled by clocking the property. In other words, while two tools might give a different result for a given design on an unclocked property, using a clocked property should guarantee the same result. In particular, using a clocked property should guarantee that a cycle-based simulator and an event-based simulator will give identical results for a given design and property.

Sampling semantics (when the value of a signal should be sampled during simulation) are not dictated by PSL for the same reason that the granularity of time is not. PSL specifies that the evaluation of a property should not influence the results of a verification – thus sampling semantics must be determined by the tool, not by PSL.

6.4 Placement of the clock

In all of our examples so far, we have shown the clock applied to the operand of the `always` operator, as in Property 6.10a, rather than to the entire property, as in Property 6.10b. The two placements of the clock are equivalent, therefore you do not need to worry about the difference – there is none! However, it may be instructive to understand why the equivalence holds.

Property 6.10b applies the clock to the entire property `always a`. Thus, it holds on Trace 6.10(i) if property `always a` holds on the trace formed from the shaded cycles of Trace 6.10(ii). Property 6.10a applies the `always` operator to the sub-property `a@(posedge clk)`. Thus, it holds if sub-property `a@(posedge clk)` holds on every cycle of the trace.

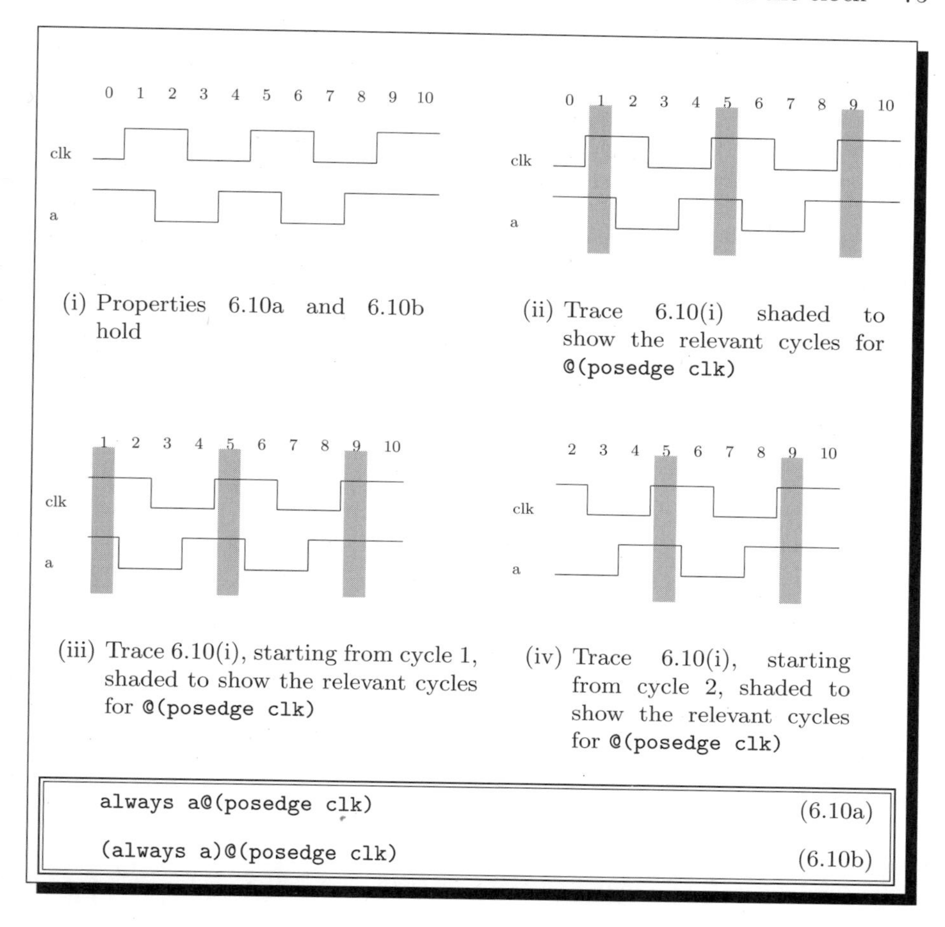

Fig. 6.10: Placement of the clock

Let's examine closely the value of sub-property `a@(posedge clk)` on cycles 0, 1, and 2. Sub-property `a@(posedge clk)` holds at cycle 0 of Trace 6.10(i) if `a` holds on the trace obtained by filtering out all cycles except those on which `posedge clk` holds. Thus, sub-property `a@(posedge clk)` holds at cycle 0 of Trace 6.10(i) if `a` holds on the shaded cycles of Trace 6.10(ii). But `a` has no temporal operator, thus it holds on the shaded cycles of Trace 6.10(ii) if it holds at the first shaded cycle.

Similarly, sub-property `a@(posedge clk)` holds at cycle 1 of Trace 6.10(i) if `a` holds on the sub-trace starting at cycle 1 obtained by filtering out all cycles except those on which `(posedge clk)` holds. Thus, sub-property `a@(posedge clk)` holds at cycle 1 of Trace 6.10(i) if `a` holds on the shaded cycles of Trace 6.10(iii). But `a` has no temporal operator, thus it holds on the shaded cycles of Trace 6.10(iii) if it holds at the first shaded cycle.

Finally, sub-property `a@(posedge clk)` holds at cycle 2 of Trace 6.10(i) if `a` holds on the sub-trace starting at cycle 2 obtained by filtering out all cycles except those on which `(posedge clk)` holds. Thus, sub-property `a@(posedge clk)` holds at cycle 2 of Trace 6.10(i) if `a` holds on the shaded cycles of Trace 6.10(iv). But `a` has no temporal operator, thus it holds on the shaded cycles of Trace 6.10(iv) if it holds at the first shaded cycle.

Note that the first shaded cycle when evaluating `a@(posedge clk)` on cycle 0 of Trace 6.10(ii) is the same as the first shaded cycle when evaluating `a@(posedge clk)` on cycle 1 of Trace 6.10(iii). This is the result of Property 6.10a applying the `always` first, then doing the filtering. In contrast, Property 6.10b first did the filtering, then applied the `always`. Of course, the end result is the same.

Properties 6.10a and 6.10b are equivalent, although the order in which they apply the `always` and the `@` operators is different. Do not be misled into thinking that one way of writing the property is more efficient than another! It is up to the tools to ensure that both ways of writing the property are efficiently implemented – this is something that the user should not have to worry about. The purpose of this explication has been to help you see why Properties 6.10a and 6.10b are equivalent, not to guide you into writing one way or the other.

6.5 Checking for glitches in a clocked design

In our previous discussions of clocking, we have always been careful to note that we assume that the design is glitch-free. This is because if there are glitches, then we cannot guarantee that our reasoning about which cycles can be filtered out safely is sound. Happily, we can express the fact that a design is glitch-free in PSL.

Consider an edge-triggered design in which signals are clocked on the rising edge of the clock, or a level-sensitive design in which latches are sensitive to the high value of the clock. If such designs are glitch-free, we expect signals to change value only once during a clock cycle, while the clock is high (in both cases we don't know exactly when the change will take place, as that depends on the delay of the particular latches).

For instance, signal `noglitch` is glitch-free in Trace 6.11(i), but signal `glitchy` is not, since it changes value more than once per cycle. We can express the fact that a signal should be glitch-free in PSL, as shown in Properties 6.11a and 6.11b. Both properties use the built-in function `stable()`, which holds if its operand has the same value that it had at the previous cycle. Property 6.11a holds on Trace 6.11(i), while 6.11b does not.

Note that Properties 6.11a and 6.11b are unclocked properties. We need access to every PSL cycle in order for them to work – thus we need an event-based tool that does not collapse everything that occurs between the clock ticks into a single cycle, as would happen with a cycle-based tool.

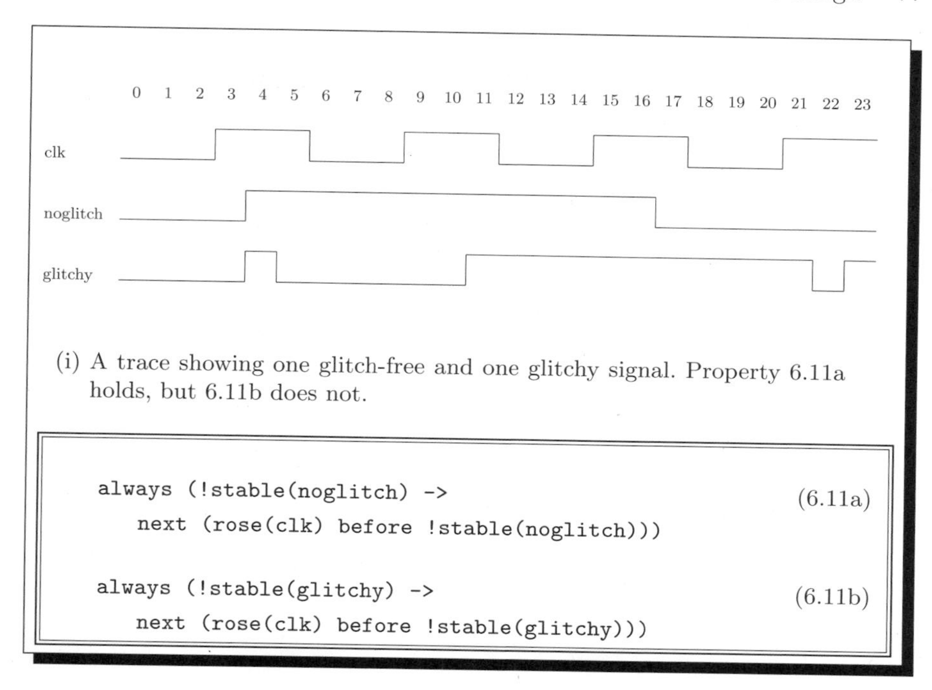

(i) A trace showing one glitch-free and one glitchy signal. Property 6.11a holds, but 6.11b does not.

```
always (!stable(noglitch) ->                                  (6.11a)
    next (rose(clk) before !stable(noglitch)))

always (!stable(glitchy) ->                                   (6.11b)
    next (rose(clk) before !stable(glitchy)))
```

Fig. 6.11: Using PSL to check for glitches

6.6 Checking for races with the clock in a clocked design

Another assumption we have made throughout is that no signal changes value at the same time as the clock, otherwise our reasoning about which cycles can be filtered out safely is not sound. We can easily express the property that a signal does not change value at the same time as the clock. Property 6.12a checks this for signal a. As with glitches, we need access to every PSL cycle in order for such a property to work – thus we need an event-based tool that does not collapse everything that occurs between the clock ticks into a single cycle, as would happen with a cycle-based tool.

```
always (rose(clk) -> stable(a));                              (6.12a)
```

Fig. 6.12: Using PSL to check for race conditions with the clock

6.7 Using the clock operator to simplify properties

PSL allows you to clock with any Boolean expression `b` – it is not necessary that `b` represent a design clock. This can be useful for simplifying properties, as described below.

6.7.1 Simplifying properties for a cycle-based trace

Consider the situation where you have a single clock and a cycle-based tool. Therefore, unclocked properties are sufficient to describe your design. Consider further that in this situation you want to express the specification "two consecutive writes will not be to the same address". One way is as shown in Assertion 6.13a. Assertion 6.13a states that if we see a write to address `i`, the next write will be to a different address, and it holds on Trace 6.13(i). We can write the assertion in a simpler manner by using the clock operator to "clock" the assertion with the write signal, as in Assertion 6.13b. The value of

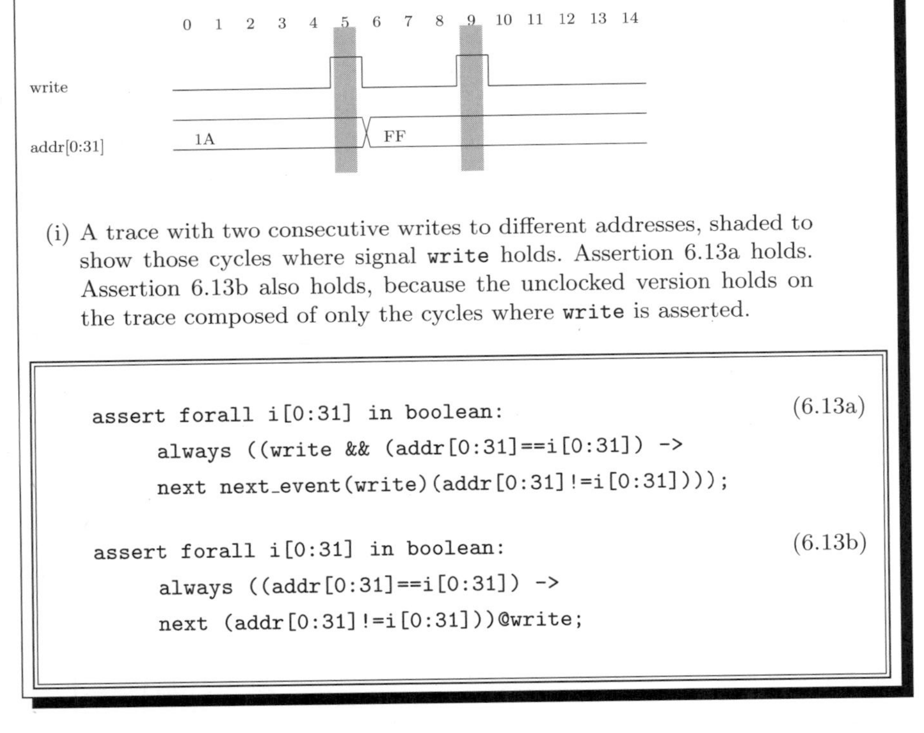

(i) A trace with two consecutive writes to different addresses, shaded to show those cycles where signal `write` holds. Assertion 6.13a holds. Assertion 6.13b also holds, because the unclocked version holds on the trace composed of only the cycles where `write` is asserted.

```
assert forall i[0:31] in boolean:                              (6.13a)
    always ((write && (addr[0:31]==i[0:31]) ->
    next next_event(write)(addr[0:31]!=i[0:31])));

assert forall i[0:31] in boolean:                              (6.13b)
    always ((addr[0:31]==i[0:31]) ->
    next (addr[0:31]!=i[0:31]))@write;
```

Fig. 6.13: Using the `@` to simplify properties on a cycle-based trace

Assertion 6.13b on a trace is the same as the value of the unclocked version on the trace formed from those cycles where write holds, shaded in Trace 6.13(i).

6.7.2 Simplifying properties for a non-cycle-based trace

For simplicity, the "trick" of Section 6.7.1 was presented for cycle-based traces. But we can use the same trick for non-cycle-based traces. For instance, in the case of a singly-clocked design with edge-triggered latches sensitive to the rising edge of the clock, we can clock as shown in Assertion 6.14a. Equivalently, we could clock the assertion with the Boolean expression `write && rose(clk)`, as shown in Assertion 6.14b. Assertion 6.14b holds on Trace 6.14(i) because its unclocked version holds on the trace formed from those cycles where `write && rose(clk)` holds, shaded in Trace 6.14(i).

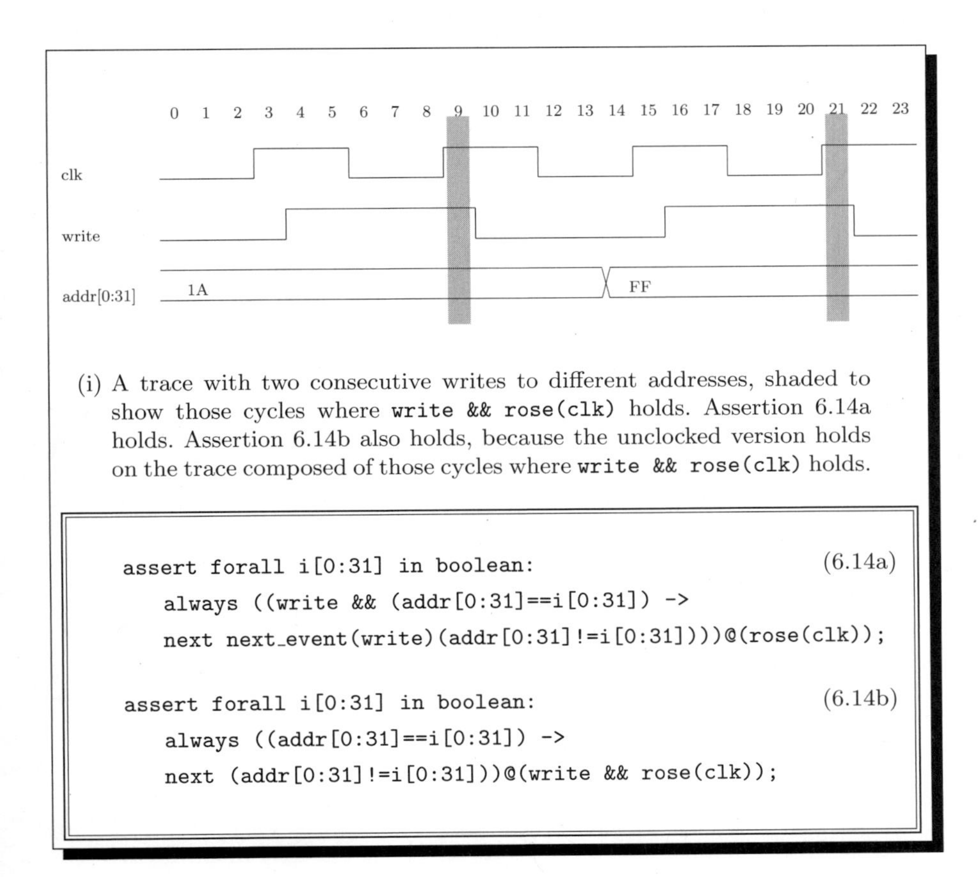

(i) A trace with two consecutive writes to different addresses, shaded to show those cycles where `write && rose(clk)` holds. Assertion 6.14a holds. Assertion 6.14b also holds, because the unclocked version holds on the trace composed of those cycles where `write && rose(clk)` holds.

```
assert forall i[0:31] in boolean:                                    (6.14a)
   always ((write && (addr[0:31]==i[0:31]) ->
   next next_event(write)(addr[0:31]!=i[0:31])))@(rose(clk));

assert forall i[0:31] in boolean:                                    (6.14b)
   always ((addr[0:31]==i[0:31]) ->
   next (addr[0:31]!=i[0:31]))@(write && rose(clk));
```

Fig. 6.14: Using @ to simplify properties on a non-cycle-based trace

6.8 The default clock

PSL provides a way to declare a default clock, so that the clock operator does not need to be used in every single property. The default clock declaration is part of the verification layer, and is used in a vunit as shown in Figure 6.15. The default clock `(posedge clk)` of Figure 6.15 is applied to every verification directive (in this case, two assertions) of the vunit. Thus, vunit `default_clock_example` of Figure 6.15 is equivalent to the vunit with the same assertions clocked on `@(posedge clk)` as shown in Figure 6.16.

```
vunit default_clock_example {

    default clock = (posedge clk);

    assert always (a || b);
    assert always (req -> next_e[1:3] (ack));

}
```

Fig. 6.15: Example of the declaration of a default clock in a vunit

```
vunit explicit_clock {

    assert always  (a || b)@(posedge clk);
    assert always (req -> next_e[1:3] (ack))@(posedge clk);

}
```

Fig. 6.16: A vunit equivalent to that of Figure 6.15

NOTE: Technically, the default clock is applied to the whole property rather than to the operand of the `always` operator, as shown in Figure 6.16. However, the property `always p@(posedge clk)` is equivalent to `(always p)@(posedge clk)`, thus for clarity we've chosen to show the equivalent `vunit` without parenthesizing the asserted properties.

Clocks do not accumulate. Thus, if we explicitly clock a property inside of a vunit which has a default clock, the explicit clock takes precedence over the default clock. For instance, vunit `default_and_explicit_clock` of Figure 6.17 is equivalent to vunit `clocks_do_not_accumulate` of Figure 6.18.

```
vunit default_and_explicit_clock {

   default clock = (posedge clk1);

   assert always  (a || b);
   assert always (req -> next_e[1:3] (ack))@(posedge clk2);

}
```

Fig. 6.17: An explicitly clocked property inside of a vunit with a default clock

```
vunit clocks_do_not_accumulate {

   assert always  (a || b)@(posedge clk1);
   assert always (req -> next_e[1:3] (ack))@(posedge clk2);

}
```

Fig. 6.18: A vunit equivalent to that of Figure 6.17

6.9 Embedding asynchronous properties

Suppose that we want to embed an asynchronous property or sequence inside of a clocked property or sequence. We can do this by "protecting" the asynchronous property or sequence by clocking it with `'true`. For instance, if we would like to say that if `a` happens on an edge of the clock, then starting at the next cycle we should see a sequence of cycles in which `b` happens on an edge of the clock, then an interrupt (assertion of signal `int`) happens asynchronously, then `c` happens on an edge of the clock, we can do it as shown in Assertion 6.19a. The `@(posedge clk)` is applied to the entire property. However, the `@'true` on `int[->]` "protects" it from the influence of the clock, and ensures that the interrupt will be interpreted asynchronously. To understand this, recall that clocks do not accumulate. That is, when clocks are nested, an inner clock takes precedence over an outer clock. In this case `'true` is func-

```
assert always ({a} |=>                                    (6.19a)
      {b ; int[->]@'true ; c})@(posedge clk);
```

Fig. 6.19: An embedded asynchronous property

tioning syntactically as a clock. But because it “ticks” every PSL cycle, the effect is to identify assertions of `int` asynchronously.

7

Aborting a Property

The abort operator is used to describe reset functionality. For example, consider the case that we want to state that every request is followed by an acknowledge, except that if a reset occurs, we are not required to see the acknowledge. Assuming that a request is indicated by an assertion of signal req, an acknowledge by an assertion of ack, and a reset by an assertion of rst, we could use Property 7.1a to express our specification. Property 7.1a holds on Trace 7.1(i) because every request is eventually followed by an acknowledge, except for those requests that are aborted by a reset.

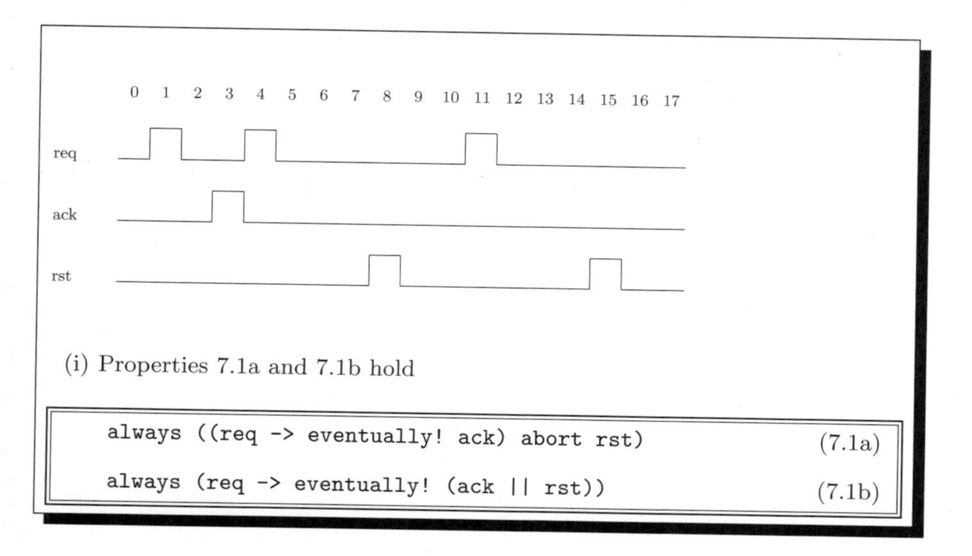

Fig. 7.1: The abort operator

NOTE: The right-hand operand of an `abort` operator (the *abort condition*) must be a Boolean expression.

In our simple example, the utility of the `abort` operator is not immediately apparent because we could have written the same property without it, as shown in Property 7.1b. However, it is not always easy to do the same for other properties. For example, consider Property 7.2a. If we want to indicate that an assertion of `rst` cancels the right-hand side, we would have to break the right-hand side into pieces, as shown in Property 7.2b. The `abort` operator allows us to express Property 7.2b in a much simpler manner, as shown in Property 7.2c. The idea behind the `abort` operator is that up until the assertion of the abort condition, everything must "go according to plan". But as soon as the abort condition is seen, any behavior can be exhibited by the trace. For example, Property 7.2c holds on Trace 7.2(i) because up until the assertion of signal `rst` at cycle 9, nothing had gone wrong: we had seen the start of a sequence of cycles that seemed to be headed towards satisfaction of the right-hand side.

Notice the way we have parenthesized Property 7.2c. The `abort` operator is applied to the sub-property that is an operand of the `always` operator in the original, unaborted property, and not to the whole property including the `always` operator. By doing so, we are aborting each particular "instance" of the sub-property created by the `always` operator.

NOTE: Recall that the use of the term "instance" does not imply that any actual instantiation is taking place, or that multiple instances are being created by a tool. Rather, the term "instance" is a conceptual term that is useful in understanding the purpose of the `always` operator.

Getting back to our point about placement of the parentheses, if we had aborted the entire property including the `always` operator (as in Property 7.3a), then we would obtain a property that does not "restart itself" after an assertion of `rst`. Aborting the entire property as in Property 7.3a means that as long as nothing went wrong up to the first assertion of the abort condition, the property will hold. For example, Property 7.3a holds on Trace 7.3(i) because the assertion of signal `rst` at cycle 9 aborts the entire property, thus the second set of cycles on which `{a;b;c}` holds does not need to see the right-hand side `{d ; e[+] ; f[+] ; g}`. But Property 7.3b does not hold on Trace 7.3(i) because the placement of the `always` operator means that the sub-property `({a;b;c} |=> {d ; e[+] ; f[+] ; g}) abort rst` must hold at all cycles, including for example at cycle 12, when the second set of cycles matching `{a;b;c}` starts.

Because the usual purpose of a reset signal in a design is not to give blanket approval for all that comes after, the normal placement of the `abort` operator is such that it is applied to the operand of the `always` operator, rather than to the entire property. What about properties whose outermost operator is not an `always`? How should they be aborted? The answer, of course, depends on what you are trying to express using PSL, and on the particular properties being aborted. However, we shall try to give a few common examples below.

0 1 2 3 4 5 6 7 8 9 10 11 12 13 14

a

b

c

d

e

f

g

rst

(i) Properties 7.2b and 7.2c hold, but 7.2a does not

```
always {a;b;c} |=> {d ; e[+] ; f[+] ; g}                (7.2a)

always {a;b;c} |=>                                       (7.2b)
      {{rst} |
      {d;rst} |
      {d ; e[+] ; rst} |
      {d ; e[+] ; f[+] ; rst} |
      {d ; e[+] ; f[+] ; g}}

always (({a;b;c} |=>                                     (7.2c)
      {d ; e[+] ; f[+] ; g}) abort rst)
```

Fig. 7.2: Aborting a non-trivial property

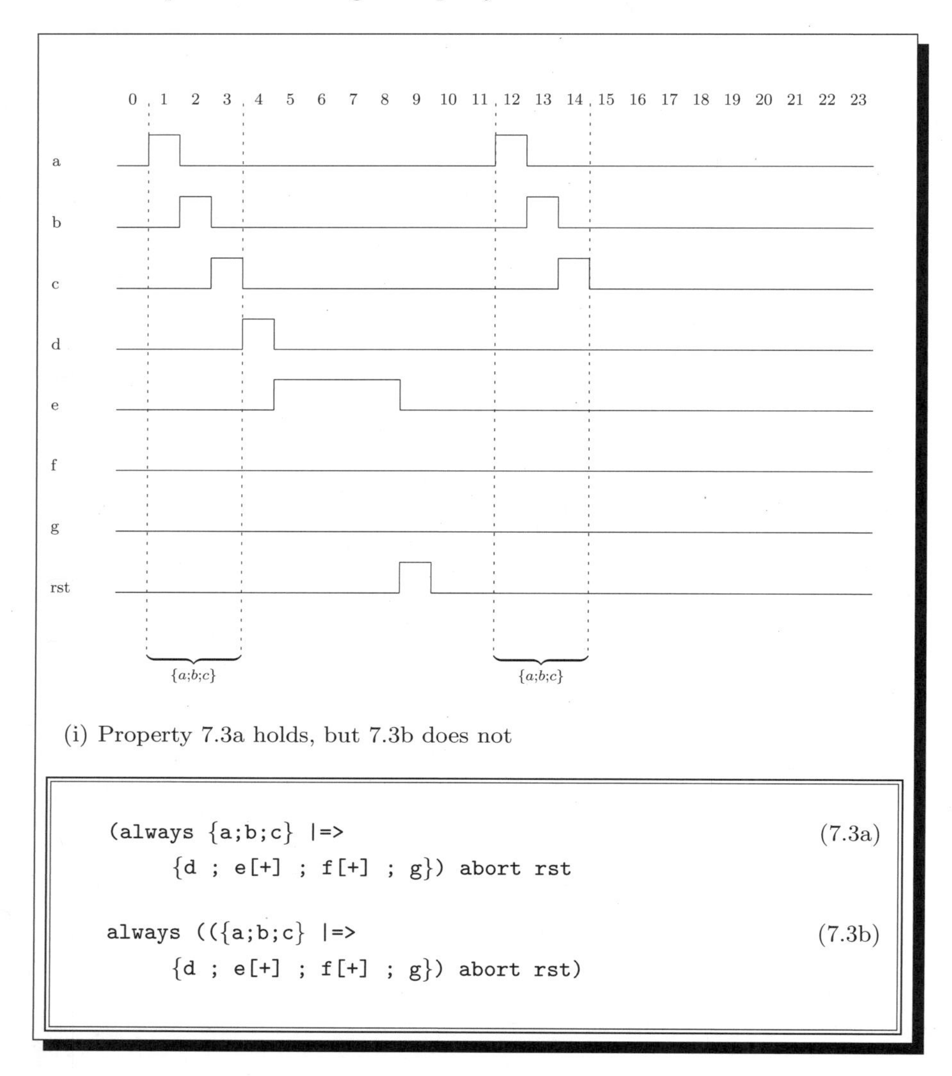

(i) Property 7.3a holds, but 7.3b does not

```
(always {a;b;c} |=>                                    (7.3a)
      {d ; e[+] ; f[+] ; g}) abort rst

always (({a;b;c} |=>                                   (7.3b)
      {d ; e[+] ; f[+] ; g}) abort rst)
```

Fig. 7.3: Placement of parentheses for the `abort` operator

7.1 Properties whose outermost operator is not `always`

If your property begins with a `never` operator, as in Property 7.4a, and you wish to apply an abort on assertion of signal `rst`, the simplest option is to notice that Property 7.4a is equivalent to Property 7.4b. Therefore, Property 7.4a can be aborted with `rst` as shown in Property 7.4c.

If your property does not begin with one of `always` or `never`, it probably begins with a Boolean expression and includes a logical implication and an

```
never {a ; b[*] ; c [+] ; d}                                  (7.4a)

always never {a ; b[*] ; c [+] ; d}                           (7.4b)

always ((never {a ; b[*] ; c [+] ; d}) abort rst)             (7.4c)
```

Fig. 7.4: Applying `abort` to a property starting with `never`

`always` operator, as in Property 7.5a. That is, some Boolean expression at the start of the trace chooses whether or not the right-hand side of the implication must hold in order for the entire property to hold. In this case, in order to apply an abort on assertion of `rst`, simply abort the right-hand side of the property as you would for any property whose outermost operator is an `always`. Thus, abort Property 7.5a as shown in Property 7.5b.

```
(mode==slowmode) -> always (req -> next_e[4:7] (ack))         (7.5a)


(mode==slowmode) ->                                           (7.5b)
      always ((req -> next_e[4:7] (ack)) abort rst)
```

Fig. 7.5: Applying `abort` to a property of the form `b -> always (...)`

Another possibility is that your traces are such that the first few cycles should be ignored. For instance, you might have something similar to Property 7.6a. In this case as with the previous one, simply abort the right-hand side of the property as you would for any property whose outermost operator is an `always`. Thus, abort Property 7.6a as shown in Property 7.6b.

Another common possibility is that your property ignores the first few cycles by examining the reset signal, as in Property 7.7a. That is, wait until signal `rst` has fallen, then check the sub-property `always (req -> next_e[1:4]`

```
next[5] (always (req -> next_event(ack)(next gnt)))           (7.6a)


next[5] (always                                               (7.6b)
      ((req -> next_event(ack)(next gnt)) abort rst))
```

Fig. 7.6: Applying `abort` to a property of the form `next[n] always (...)`

```
next_event(!rst)(always (req -> next_e[1:4] (ack)))          (7.7a)

next_event(!rst)                                              (7.7b)
      (always ((req -> next_e[1:4] (ack)) abort rst))

always ((req -> next_e[1:4] (ack)) abort rst)                 (7.7c)
```

Fig. 7.7: Aborting a property of the form `next_event(...)(always(...))`

`(ack))`. In this case, there is no reason to continue to use the `next_event` operator when aborting with `rst`. That is, adding an abort on signal `rst`, as in Property 7.7b, is equivalent to the much simpler Property 7.7c. To understand this, observe that the purpose of the `next_event(!rst)` in Property 7.7a is to jump to the first cycle in which signal `rst` is not asserted, and then to begin evaluation. Thus, it takes care of those assertions of `rst` that happen up until the first deassertion of `rst`. However, aborting with `rst` takes care of *all* assertions of `rst`. Thus the use of `next_event(!rst)` in Property 7.7b is redundant, and can be removed.

Finally, your property may be about an initial condition. For instance, to express the property that requests (assertion of `req`) and grants (assertion of `gnt`) are interleaved, we might write three properties: that the very first assertion of either `req` or `gnt` is an assertion of `req`, that after every request we see a grant before another request, and that after every grant we see a request before another grant. The second and third of these properties use an `always` operator and thus it is clear how to abort them.

Let's consider the first in more detail. It appears as Property 7.8a. Property 7.8a contains neither a `never` nor an `always` operator. To abort it, we first add an `abort` operator, to get `(req before gnt) abort rst`. However, that is not enough. It takes care of aborting the property, but it does not take care of restarting the property after an assertion of `rst`. We need to do this because after a reset (assertion of `rst` in this case), we expect to see the same behavior as that seen at the very beginning of a trace. We accomplish this as shown in Property 7.8b by adding an `always` operator and an implication the left-hand side of which is the reset signal (`rst`). Finally,

```
req before gnt                                                (7.8a)

always (rst -> next ((req before gnt) abort rst))             (7.8b)
```

Fig. 7.8: Aborting a property containing neither a `never` nor an `always` operator

we need a **next** operator to ensure that the right-hand side of the implication is not aborted by the same assertion of `rst` that triggered the implication.

7.2 Synchronous vs. asynchronous abort

In the presence of clocks, PSL distinguishes between a synchronous abort, which recognizes the abort condition only if it occurs on a clock cycle, and an asynchronous abort, which recognizes abort conditions that may occur between clock cycles as well. The `async_abort` operator is asynchronous, and the `sync_abort` operator is synchronous. The plain `abort` operator, which we have seen previously, is a synonym for `async_abort`. Thus, Assertions 7.9a and 7.9b hold on Trace 7.9(i) because the asynchronous assertion of `reset` is recognized, but Assertion 7.9c does not because the asynchronous assertion of `reset` is not recognized.

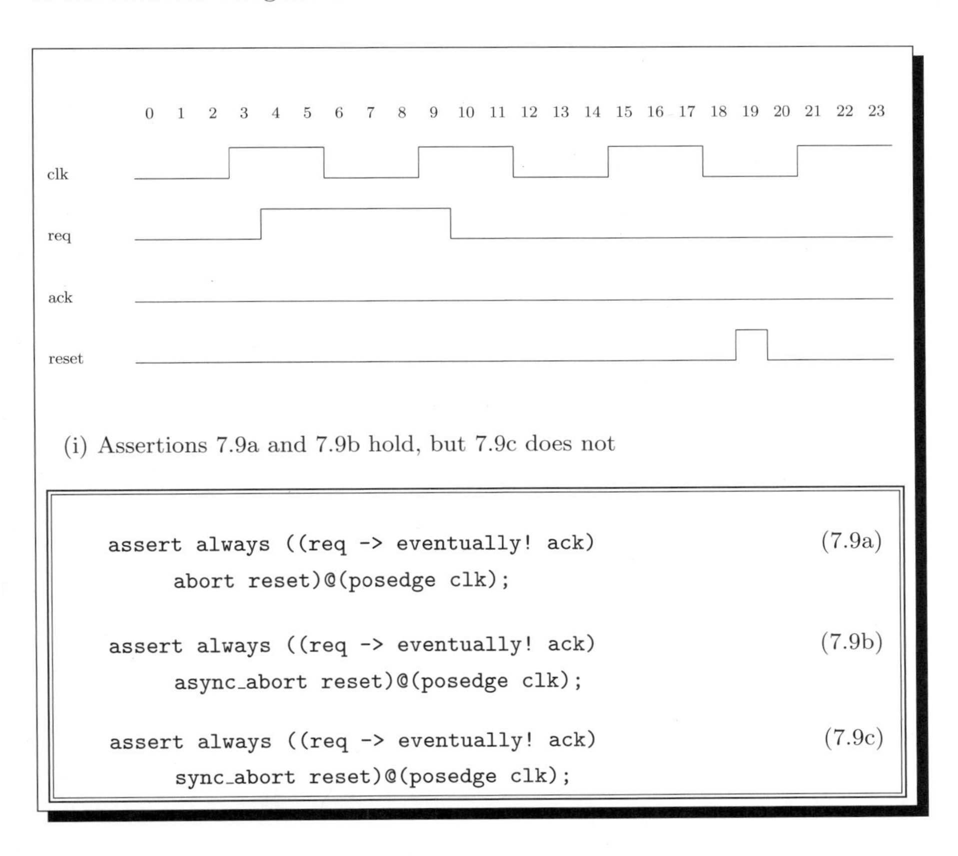

Fig. 7.9: Asynchronous vs. synchronous `abort`

8

Some Convenient Constructs

8.1 Comments

PSL comments are flavor dependent. For each flavor, the comment capability is consistent with that provided by the corresponding HDL environment. For the SystemC, SystemVerilog, and Verilog flavors, both the block comment style (`/* ....*/`) and the trailing comment style (`// ....`<`eol`>) are supported. For the VHDL flavor, the trailing comment style (`-- ....`<`eol`>) is supported. For the GDL flavor, both the block comment style (`/* ....*/`) and the trailing comment style (`-- ....`<`eol`>) are supported.

8.2 Named properties and SEREs

PSL provides a way to gives names to properties and to SEREs which can then be instantiated. For example, Figure 8.1 declares a property named `getgrant` and instantiates it twice. Property `getgrant` takes four parameters: a Boolean expression `req`, a numeric expression `n`, and two additional Boolean expressions `gnt` and `clk`. The two instantiations of `getgrant` shown in Figure 8.1 are equivalent to the two assertions shown in Figure 8.2.

Figure 8.3 shows the declaration and instantiation of two named SEREs. A named SERE is introduced by the keyword `sequence`. The instantiations of named SEREs `reqseq` and `dataseq` in Figure 8.3 result in an assertion equivalent to that of Figure 8.4.

The syntax of property and SERE declaration is flavor dependent. In the Verilog, SystemVerilog, SystemC and GDL flavors, it is as shown in Figures 8.1 and 8.3. In the VHDL flavor, the word `is` replaces the equals sign.

An entire property or named SERE can be passed as a parameter. Figure 8.5 shows a named SERE that takes another SERE and a Boolean expression as parameters, and a property that takes a SERE and a property as parameters. The instantiations shown in Figure 8.5 result in a property equivalent to that shown in Figure 8.6.

```
property getgrant (boolean req; numeric n; boolean gnt, clk) =
                always (req -> next[n] (gnt)) @ (posedge clk);

assert getgrant(req1, 3, gnt1, clk1);
assert getgrant(req2, 5, gnt2, clk2);
```

Fig. 8.1: The declaration and instantiation of a named property

```
assert always (req1 -> next[3] (gnt1)) @ (posedge clk1);
assert always (req2 -> next[5] (gnt2)) @ (posedge clk2);
```

Fig. 8.2: Two assertions equivalent to those shown in Figure 8.1

```
sequence reqseq(boolean req,ack,cancel,gnt) =
                {req ; ack & !cancel ; gnt};
sequence dataseq(boolean start,data,done; numeric n) =
                {start ; data[*n] ; done};

assert reqseq(req_out, ack_in, cancel_out, gnt_in) |=>
                   dataseq(start_data, data, end_data, 8);
```

Fig. 8.3: The declaration and instantiation of two named SEREs

```
assert {req_out ; ack_in & !cancel_out ; gnt_in} |=>
                 {start_data ; data[*8] ; end_data};
```

Fig. 8.4: An assertion equivalent to that shown in Figure 8.3

```
sequence ahead_of(sequence s; boolean fin) =
                              {{s && fin[=0]} ; fin};
property seqimplies(sequence s; property p) =
                              always {s} |=> (p);

assert seqimplies(ahead_of({a;b;c}[*],d), e before f);
```

Fig. 8.5: Passing properties and SEREs as parameters

```
assert always {{{a;b;c}[*] && d[=0]} ; d} |=> (e before f);
```

Fig. 8.6: An assertion equivalent to that shown in Figure 8.5

The formal parameters of a named property or SERE can be one of the following types:

- `boolean`: A Boolean expression
- `bit`: A single bit
- `bitvector`: A vector composed of bits
- `numeric`: Any expression interpretable as an integer in the underlying flavor. For example, in the Verilog flavor, any bitvector that does not contain an unknown bit is interpretable as a numeric.
- `string`: A string
- `sequence`: A PSL sequence. The term *sequence* is used in PSL to denote one of the following: a braced SERE, a clocked SERE, a repeated SERE, or an instantiation of a named SERE.
- `property`: A PSL property

In addition, data types of the underlying flavor can be used as formal parameter types, by prepending the keyword `hdltype` to the type. For instance, Figure 8.7 shows the declaration of a named SERE with a single parameter of type `MYTYPE`, where `MYTYPE` is a type belonging to the underlying flavor.

```
sequence type_example (hdltype MYTYPE a) =
                {a==‘NULL ; (a!=‘NULL)[*] ; a==‘NULL};
```

Fig. 8.7: Using data types of the underlying flavor

A formal parameter that is not of type `sequence` or `property` can be modified with the keyword `const`, to indicate that the actual expression that maps to the formal parameter must be statically computable. For instance, Figure 8.8 shows the declaration of a named SERE with two parameters, both of type `MYTYPE`. Parameter `b` is modified with keyword `const`, thus any call to `const_example` must pass a statically computable actual parameter for `b`.

If no type is specified for a `const` formal parameter, `numeric` is understood.

```
sequence const_example
        (hdltype MYTYPE a; const hdltype MYTYPE b) =
                {a==b ; (a!=b)[*] ; a==b};
```

Fig. 8.8: Modifying a formal parameter with `const`

8.3 The `forall` operator

The `forall` operator provides a way to write families of properties. Previously we saw a use of the `forall` operator in Assertion 2.6a, repeated for convenience in Assertion 8.9a. Assertion 8.9a states that whenever a read request (assertion of signal `read_request`) occurs, then the next four data transfers (assertion of signal `data`) must have a matching tag.

```
assert forall i in {0:7}:                                        (8.9a)
      always ((read_request && tag[2:0]==i) ->
      next_event_a(data)[1:4](tag[2:0]==i));

always ((read_request && tag[2:0]==3) ->                         (8.9b)
      next_event_a(data)[1:4](tag[2:0]==3))
```

Fig. 8.9: The `forall` operator

A property that uses the `forall` operator in PSL is known as a *replicated property*, because the same functionality can be achieved by replicating the base property a number of times and "anding" the results together. For instance, Assertion 8.9a is equivalent to the assertion of the "and" of eight separate properties, each of which refers to a specific value of `i`. One of those properties, for the value `i==3`, is shown in Property 8.9b. This does not imply that an implementation of PSL must actually perform a replication. Rather, the term "replicated property" is used figuratively, and hints at the meaning of such properties, rather than at what a tool must do to implement them.

A `forall` property can be used to link the values of two signals separated by time, as in Assertion 8.9a above. Used in this way, the `forall` operator can be thought of as giving you a way to sample the value of a signal in order to use it later on in a property. For instance, Assertion 8.9a above can be understood as sampling signal `tag[2:0]` into `i` whenever there is a read request, then comparing that sampled value to the value in `tag[2:0]` later on, when signal `data` is asserted.

Understanding `forall` in this way is useful, and indeed it is quite possible that an implementation might use some sort of sampling in order to implement the `forall` operator. However, remember that syntactically, an index variable such as `i` in Assertion 8.9a is not assigned. For instance, in Assertion 8.9a, the index `i` is used twice, both times in the Boolean expression `tag[2:0]==i`, but it is never explicitly assigned a value.

The index variable `i` of Assertion 8.9a is an integer. PSL also allows bit vectors. A classic example of the use of bit vectors in this way is shown

```
assert forall i in {0:7}:                                          (8.10a)
    forall j[127:0] in boolean:
    always ((reqin && tag[2:0]==i && dat[127:0]==j[127:0]) ->
    next_event(reqout && tag[2:0]==i)(dat[127:0]==j[127:0]));
```

Fig. 8.10: A "scoreboard" example

in Assertion 8.10a. Assertion 8.10a is a "scoreboard" assertion – for every input request (assertion of `reqin`), we expect the associated output request (assertion of `reqout` with the same tag) to have the data (`dat[127:0]`) that was read in for that tag. Assertion 8.10a can check a trace in which there are multiple requests outstanding (each associated with a different tag). Thus, it is a good illustration of the fact that while a `forall` property can be understood as sampling, something a little bit more sophisticated is going on.

We have seen that a replicated property is not necessarily implemented by replication, and that while a common use of a replicated property is to "sample" the value of a signal for later use, no sampling is necessarily taking place. In both cases, the reason is that PSL is a language for describing *what*, but not *how*. A tool may implement an operator in any way it sees fit, as long as it achieves the defined functionality, and in particular, it may implement `forall` by replicating or by sampling or in some completely different way.

8.4 Parameterized properties and SEREs

Another way to write families of properties is through parameterization of the "and" and "or" operators. Parameterization of the "and"/"or" operators provides a shortcut way to "and"/"or" together a group of properties. For instance, Property 8.11a is equivalent to applying `&&` to the set containing the four FL properties: `P(0)`, `P(1)`, `P(2)`, and `P(3)`. That is, it is equivalent to the FL Property 8.11b. The "or" operator (`||`) can be used similarly.

```
for i in {0:3}: && (P(i))                                          (8.11a)

P(0) && P(1) && P(2) && P(3)                                       (8.11b)
```

Fig. 8.11: Parameterized properties

```
assert always                                                    (8.12a)
    ((for i in {0:31}: || (data[i]!=expected[i])) ->
     data_corrupted);
```

Fig. 8.12: An example of the use of a parameterized property

Assertion 8.12a is an example of the use of a parameterized property. It states that if any of the data bits are unexpected (`data[i]!=expected[i]`), then signal `data_corrupted` must be asserted.

Parameterized SEREs can be created in a manner similar to that of parameterized properties. For instance, SERE 8.13a is equivalent to applying `&&` to the set containing the four compound SEREs: `S(0)`, `S(1)`, `S(2)` and `S(3)`. That is, it is equivalent to the compound SERE 8.13b. The SERE "or" operator (`|`) and the SERE non-length-matching "and" operator (`&`) can be used similarly.

```
for i in {0:3}: && S(i)                                          (8.13a)

{S(0) && S(1) && S(2) && S(3)}                                   (8.13b)
```

Fig. 8.13: Parameterized SEREs

Assertion 8.14a shows an example of the use of a parameterized SERE. It states that if there is a global print request (assertion of `global_print_req`), then we should subsequently see prints on each of the printers (assertion of `prt_done[i]`), followed by an indication of success (assertion of `print_success`) immediately after the last print has completed.

```
assert always {global_print_req} |->                             (8.14a)
    {{for i in {0:7}: & {prt_done[i]}[->]} ; print_success};
```

Fig. 8.14: An example of the use of parameterized SEREs

8.5 Macros

PSL macro support is flavor dependent. The Verilog and SystemVerilog flavors support Verilog compiler directives such as `‘define`, `‘ifdef`, etc. The VHDL, SystemC and GDL flavors support cpp preprocessing directives such as `#define`, `#ifdef`, etc. In addition, all flavors support PSL macros `%for` and `%if`.

The `%for` construct replicates a piece of text a number of times. The syntax is as shown in Figure 8.15 where `var` is an identifier, `expr1` and `expr2` are statically computable expressions, and `item1`, `item2` etc. are either a number or an identifier.

```
// using a range
%for var in expr1 .. expr2 do
...
%end

// using a list
%for var in { item1 , item2  , ... , itemN } do
...
%end
```

Fig. 8.15: Syntax of `%for`

In the first case the text inside the `%for...%end` pairs will be replicated `expr2-expr1+1` times (assuming that `expr2>=expr1`). In the second case the text will be replicated according to the number of items in the list. During each replication of the text, the loop variable value can be substituted into the text in one of the following two ways:

1. Given a loop variable `ii`, the current value of the loop variable can be accessed using simply `ii` if `ii` is a separate token in the text.
2. If `ii` is part of an identifier and/or part of an expression, it can be accessed using `%{<expr>}`.

For instance, the code shown in Figure 8.16 expands to the code shown in Figure 8.17.

The following operators can be used in preprocessor expressions:

```
=  !=  <  >  <=  >=  -  +  *  /  %
```

The `%if` construct is similar to the `#if` construct of the cpp preprocessor. However `%if` must be used when the condition refers to variables defined in an encapsulating `%for`. The syntax, with and without the else-part, is shown in Figure 8.18.

```
%for ii in 0..3 do
assign aa[ii] = ii > 2;
%end

%for ii in 1..4 do
assign aa%{ii-1} = bb%{ii} > 2;
%end
```

Fig. 8.16: Two ways to use the loop variable `ii`

```
assign aa[0] = 0 > 2;
assign aa[1] = 1 > 2;
assign aa[2] = 2 > 2;
assign aa[3] = 3 > 2;

assign aa0 = b1 > 2;
assign aa1 = b2 > 2;
assign aa2 = b3 > 2;
assign aa3 = b4 > 2;
```

Fig. 8.17: Expansion of the code shown in Figure 8.16

```
%if <expr> %then
...
%end

%if <expr> %then
...
%else
...
%end
```

Fig. 8.18: Syntax of `%if`

As an example of using `%for` in conjunction with `%if`, consider Figure 8.19, which checks that each of the signals `read_byte`, `read_word`, `write_byte` and `write_word` is mutually exclusive to the others.

```
%for s1 in {read_byte,read_word,write_byte,write_word} do
%for s2 in {read_byte,read_word,write_byte,write_word} do
%if (s1!=s2) %then
  assert never (s1 && s2);
%end
%end
%end
```

Fig. 8.19: An example of the use of `%for`

9

The Simple Subset

Previously, in Section 3.4, we saw Assertion 3.3a (repeated here as Assertion 9.1a), in which, if `a` is asserted at cycle N, we must look ahead to cycle $N + 6$ to see whether `b` holds in order to know whether or not we require `c` to be asserted at N and `d` to be asserted at $N + 2$. Such an assertion is not a part of the *simple subset* of PSL. In the simple subset, time flows from left to right through the property, in the following way: considering each Boolean expression and each SERE as an atomic entity, if we need to know the value of such an entity at cycle M, then the value of everything to the right of the entity need not be known before cycle M.

```
assert always ((a && next[6](b)) -> (c && next[2](d)));   (9.1a)
```

Fig. 9.1: A property not in the simple subset

For a PSL property to be in the simple subset, it must obey the rules shown in Table 9.1.[1] All operators not appearing in the table, and in particular the operators `always`, `next`, `next_a`, `next_event`, `next_event_a`, and all forms of suffix implication are allowed without restriction in the simple subset.

The simple subset is mostly a matter of form, not content. In other words, many properties not in the simple subset can be rewritten into the simple subset. For example, Assertion 9.2a, stating that if we see a request without an acknowledge in the next three to five cycles, then we must have seen a cancellation the cycle after the request, does not belong to the simple subset. However, the equivalent Assertion 9.2b, stating that if we see a request without

[1] The table refers to the logical iff operator, `<->`, which we have not yet seen. The `<->` operator is not widely used because of the restriction in the simple subset that both its operands be Boolean. The property `a <-> b` is equivalent to `(a -> b) && (b -> a)`.

Table 9.1: The simple subset. Here `before*` stands for all the variations of the `before` operator.

<table>
<tr><th>PSL operator</th><th>Restriction</th></tr>
<tr><td><code>!</code></td><td>Operand must be Boolean</td></tr>
<tr><td><code>never</code></td><td>Operand must be Boolean or sequence</td></tr>
<tr><td><code>eventually!</code></td><td>Operand must be Boolean or sequence</td></tr>
<tr><td><code>||</code></td><td>At most one operand may be non-Boolean</td></tr>
<tr><td><code>-></code></td><td>Left-hand side must be Boolean</td></tr>
<tr><td><code><-></code></td><td>Both operands must be Boolean</td></tr>
<tr><td><code>until</code> <code>until!</code></td><td>Right-hand side must be Boolean</td></tr>
<tr><td><code>until_</code> <code>until!_</code></td><td>Both operands must be Boolean</td></tr>
<tr><td><code>before*</code></td><td>Both operands must be Boolean</td></tr>
<tr><td><code>next_e</code></td><td>Operand must be Boolean</td></tr>
<tr><td><code>next_event_e</code></td><td>Right-hand operand must be Boolean</td></tr>
</table>

```
always ((req && !next_e[3:5] (ack)) -> (next cancel))    (9.2a)

always ({req ; !cancel} |-> {[*3:5] : ack})              (9.2b)
```

Fig. 9.2: The same property expressed two ways

a cancellation, then we had better see an acknowledge within three to five cycles, does belong to the simple subset.

Properties outside of the simple subset can be difficult to understand. In addition, implementing tools that support only the simple subset is easier than implementing tools that support all of PSL. Thus, many tools do not support properties that are outside of the simple subset of PSL. Finally, we have never come across a hardware design in which a needed property could not be expressed within the bounds of the simple subset. For all of these reasons, it is recommended to stay within the simple subset when writing properties.

10

The Boolean, Modeling, and Verification Layers

Up until now, we have focused almost exclusively on the temporal layer. In this chapter, we briefly discuss various aspects of the Boolean, modeling and verification layers not yet covered.

10.1 The Boolean layer

The Boolean layer consists of any Boolean expression in the underlying flavor. A Boolean expression is an expression that is evaluated in a single cycle, and has the value true or false. Note that, as in most languages, a Boolean expression may contain a non-Boolean expression. For instance, the Boolean expression `a[2:0] > 3'b2` contains the bit vector `a[2:0]` and the constant `3'b2`, both of them non-Boolean.

In addition, any single bit is interpretable as a Boolean expression. For instance, the single bit `a[3]` may appear anywhere that a Boolean expression is required, even if `a[3]` may have a non-Boolean value such as X or Z from the four-valued logic of Verilog. In the case that the value is non-Boolean, the bit is interpreted as true or false according to the rules of the underlying flavor, in the same way that the underlying flavor treats such an expression when it appears as the condition of an if-statement.

Finally, in the Verilog, SystemVeriog and SystemC flavors, a bit vector is also interpretable as a Boolean expression, and similarly to a single bit, is interpreted as true or false according to the rules of the underlying flavor, in the same way that the underlying flavor treats such an expression when it appears as the condition of an if-statement.

PSL provides a set of built-in functions that can be used wherever a Boolean expression is expected. In the descriptions of the built-in functions below, the term *clock expression* refers to an expression following the `@` symbol (which may be either a Boolean expression or `(posedge b)` or `(negedge b)` for some Boolean expression `b`). The term *clock context* refers to the clock expression that influences a given expression. For instance, in the property `always (a -> next b)@(posedge clk)`, the clock context of both `a` and `b` is `(posedge clk)`.

The built-in functions are:

- `prev()`: The built-in function `prev()` may be invoked using one of the following three forms: `prev(e)`, `prev(e,N)`, `prev(e,N,c)`, where `e` is an expression of any type, `N` is a number, and `c` is a clock expression. The form `prev(e)` returns the previous value of `e`, that is, the value that `e` had the previous cycle, with respect to the clock of its context. The form `prev(e,N)` returns the value of `e` in the Nth previous cycle, with respect to the clock of its context, and the form `prev(e,N,c)` returns the value of `e` in the Nth previous cycle, with respect to clock `c`.
- `next()`: The built-in function `next()` is invoked using the following form: `next(e)`, where `e` is an expression of any type. See Section 11.4.4 for an explanation of the returned value.
- `stable()`: The built-in function `stable()` may be invoked using one of the following two forms: `stable(e)`, `stable(e,c)`, where `e` is an expression of any type, and `c` is a clock expression. The form `stable(e)` returns true if `e` is the same as `prev(e)`, with respect to the clock of its context. The form `stable(e,c)` returns true if `e` is the same as `prev(e,1,c)`.
- `rose()`: The built-in function `rose()` may be invoked using one of the following two forms: `rose(b)`, `rose(b,c)`, where `b` is a Boolean expression and `c` is a clock expression. The form `rose(b)` returns true if `b` is asserted and was deasserted the previous cycle, with respect to the clock of its context. `rose(b,c)` returns true if `b` is asserted and was deasserted the previous cycle, with respect to clock `c`.
- `fell()`: The built-in function `fell()` is similar to `rose()`, but looks for a falling edge rather than a rising edge.
- `ended()`: The built-in function `ended()` may be invoked using one of the following two forms: `ended(r)`, `ended(r,c)`, where `r` is a SERE and `c` is a clock expression. The meaning of `ended(r)` for some SERE `r` was explained in Section 5.8. The form `ended(r,c)` is equivalent to `ended(r@c)`.
- `isunknown()`: For a bit vector `V`, the built-in function `isunknown(V)` returns true if `V` contains any bit with an unknown value.
- `countones()`: For a bit vector `V`, the built-in function `countones(V)` returns the number of bits of `V` that have the value 1.
- `onehot()`: For a bit vector `V`, the built-in function `onehot(V)` returns true if `V` has exactly one bit with value 1.
- `onehot0()`: For a bit vector `V`, the built-in function `onehot0(V)` returns true if `V` contains at most one bit with value 1.
- `nondet()`: The built-in function `nondet()` takes a list of values as an argument. See Section 11.4.3 for an explanation of the returned value.
- `nondet_vector()`: The built-in function `nondet_vector()` takes two arguments: a number and a list of values. See Section 11.4.3 for an explanation of the returned value.

10.2 The modeling layer

The modeling layer provides a means to model behavior of design inputs, and to declare and give behavior to auxiliary signals and variables. This part of the modeling layer is simply a subset of the underlying flavor. For the SystemVerilog, Verilog, and VHDL flavors, the modeling layer consists of the synthesizable subset of the respective flavors. The SystemC flavor of the modeling layer consists of those declarations and statements which would be legal in the context of the SystemC module to which the vunit is bound. The GDL flavor of the modeling layer consists of all of GDL.

Figure 10.1 shows a vunit that makes use of the modeling layer to declare an auxiliary signal `valid_read_request`, then uses that signal in an assertion. See Chapters 11, 12, and 13 for some additional examples of the use of the modeling layer.

```
vunit modeling_example {
   wire valid_read_request;

   assign valid_read_request = read && read_en && !busy;

   assert always (valid_read_request -> eventually!  data_valid);
}
```

Fig. 10.1: Using the modeling layer to declare an auxiliary signal

10.3 The verification layer

The verification layer includes the declaration of verification units using the keywords `vunit`, `vmode` and `vprop`, and also provides verification directives such as `assert`. We have seen both throughout this book. In this chapter, we provide a little more detail on the use of verification units and directives.

10.3.1 Verification units

The relation between signals in a verification unit and those of the design under verification is by name, relative to the module or module instance to which a verification unit is bound. For example, verification unit `binding_example` in Figure 10.2 is bound to module `mod1`, thus signals `a` and `b` refer to signals of that name as declared in module `mod1`.

A verification unit can inherit another verification unit, using the keyword `inherit`. For instance, vunit `inherit_example` of Figure 10.3 inherits vunit `some_declarations`.

```
vunit binding_example(mod1) {
   assert never (a && b);
}
```

Fig. 10.2: The vunit `binding_example` is bound to module `mod1`

```
vunit some_declarations {
   property p1(boolean a, b) = never a && b;
   property p2(boolean a, b) = always (a -> next b);
}
vunit inherit_example {
   inherit some_declarations;

   assert p1(sig1, sig2);
   assert p2(sig3, sig4);
}
```

Fig. 10.3: An example of vunit inheritance

If an inherited vunit and an inheriting vunit declare the same signal name, the name declared in the inheriting vunit takes precedence. A verification unit may override the behavior of a design signal. This is done by redeclaring a design signal name and also giving it behavior in the verification unit. In Section 11.4.5 we provide an example.

Two special types of verification units are the `vmode` and the `vprop`. A `vmode` cannot contain assert directives, while a `vprop` can contain only assert directives. Together with vunit inheritance, these special types of verification units can be used to structure a specification.

A verification unit named `default` must be of the special type `vmode`. It is automatically inherited by all other verification units, and may not itself inherit other vunits.

10.3.2 Verification directives

The verification layer also provides the verification directives, which are:

- `assert`: This is the most basic verification directive, which we have used extensively throughout this book. It instructs a verification tool to verify that a property holds.
- `assume`: This directive instructs a verification tool to constrain the verification so that a property holds. In other words, it instructs the verification tool that any asserted properties need not hold on traces that do not obey the assumptions. Assumptions will normally be made on the inputs of a design under verification.

- `assume_guarantee`: This directive instructs a verification tool to assume that a property holds, and in addition, to verify that the assumption holds. For example, if block A is the design under verification and contains assumptions about input signals driven by block B, then using `assume_guarantee p;` in block A will instruct the tool to check `assume p;` in block A and `assert p;` in block B.
- `restrict`: This directive restricts a verification tool to constrain the verification so that it matches a SERE. For instance, `restrict {a ; b ; c[*]}` instructs the tool to consider only traces every prefix of which matches `{a ; b ; c[*]}`.
- `restrict_guarantee`: Similarly to `assume_guarantee`, this directive instructs a verification tool to restrict the verification to traces matching a SERE, and in addition, to verify that the restriction holds in the driving block or blocks.
- `cover`: This directive instructs a verification tool to check if a trace matching a given SERE was seen during a verification run.
- `fairness`: This verification directive is used in formal verification. See any standard text on formal verification for the use of fairness constraints.
- `strong fairness`. Like `fairness`, this verification directive is used in formal verification. See any standard text on formal verification for the use of fairness constraints.

Verification directives can be labeled, in order to allow unique identification of PSL directives from tool interfaces and in textual reports. For example, Figure 10.4 shows a labeled assertion.

```
rule103:  assert always (req -> next_e[1:3] (ack));
```

Fig. 10.4: A labeled assertion

The keyword `report` can be used with the verification directives `assert`, `assume_guarantee`, `restrict_guarantee`, and `cover` to provide a string containing a message to report when the property fails to hold or is covered. For example, Figure 10.5 shows an assertion that prints "`a and b were asserted together`" whenever a violation of the property `never (a && b)` is seen.

```
assert never (a && b) report "a and b were asserted together";
```

Fig. 10.5: The use of `report`

11

Advanced Topics

11.1 Finite traces – the four levels of satisfaction

Until now, we have spoken in terms of a PSL property holding or not on a given trace, and in fact, the meaning of a PSL operator is given by defining when a property using it holds or does not hold. However, PSL actually defines four additional terms that can be used to relate a property to a trace. The four levels of satisfaction are:

1. holds strongly
2. holds but does not hold strongly
3. pending
4. fails

All four terms are based upon the basic notion of a property holding, and all four refer to what would happen to a finite trace if the simulation were to be continued, thus lengthening the trace. Would the property continue to hold or not? *Holds strongly* means that no matter what we do, we cannot undo the holding of the property. *Holds but does not hold strongly* means what it says: the property holds, but not strongly. *Pending* means that while the property doesn't hold, continuing the simulation could go either way, depending on how we lengthen the trace. *Fails* means that the property doesn't hold, and in addition, we cannot get it to hold by lengthening the trace. Thus, *holds strongly* and *holds but does not hold strongly* are two degrees of holding, while *pending* and *fails* are two degrees of not holding.

NOTE: Similarly to the leniency and strictness of strong and weak operators discussed in Chapter 4, the four degrees of holding have no opinion on the Boolean expressions `'true` and `'false`. Thus, the property `a -> next![3] ('true)` does not hold strongly on a trace two cycles long where `a` holds on the first cycle, because `'true` is not regarded as a sure thing. And the property `a -> next[3] ('false)` is pending on the same trace, because `'false` is not viewed as a lost cause.

Property 11.1a holds strongly on Trace 11.1(i). On Trace 11.1(ii) it is pend-

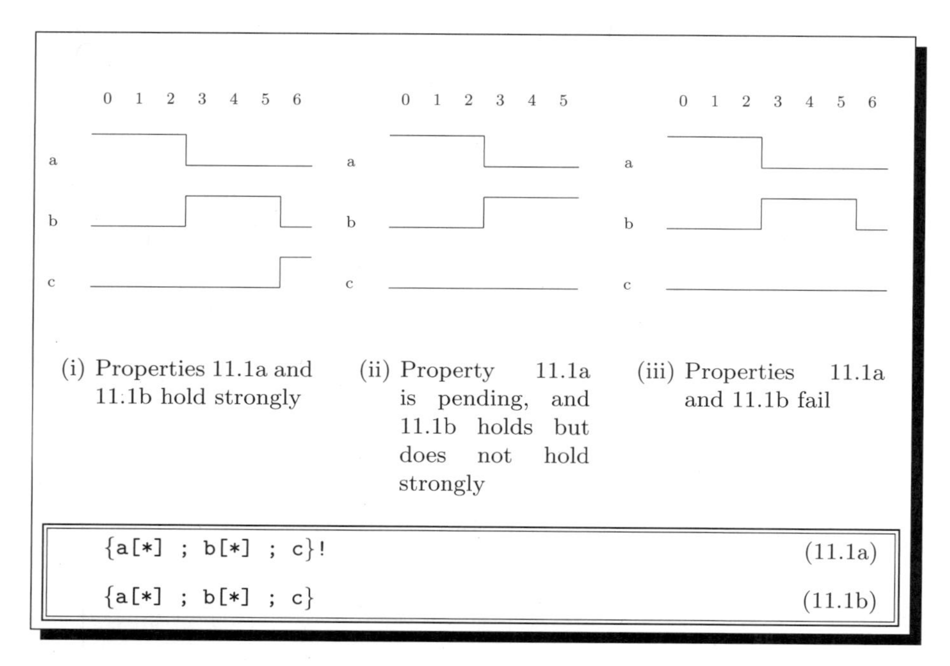

Fig. 11.1: A weak vs. strong SERE

ing because we haven't yet seen `c`, but depending on how the trace continues, the property may end up holding on it. It fails on Trace 11.1(iii) because no matter how we continue the trace, it cannot be "saved". Property 11.1b holds strongly on Trace 11.1(i). It holds on Trace 11.1(ii), but does not hold strongly on this trace, because we do not know what will happen if the trace is extended. Property 11.1b fails on Trace 11.1(iii).

Thus the difference between Properties 11.1a and 11.1b on the traces of Figure 11.1 is evident only on Trace 11.1(ii). Property 11.1a is pending on this trace, while Property 11.1b holds on it. A weak property such as Property 11.1b is never pending, because if such a property does not hold, then extending the trace cannot make it hold. Conversely, a strong property such as Property 11.1a never holds but does not hold strongly, since the strength of the operator prohibits such a situation.

However, a property that contains a mixture of strong and weak operators can be pending and can hold but not hold strongly. For instance, Property 11.2a is pending on Trace 11.2(i) because extending the simulation may reveal a cycle where `b` holds, which is needed because of the assertion of `a` at cycle 6. It holds but does not hold strongly on Trace 11.2(ii), because extending the simulation may reveal another cycle where `a` holds without a future `b`.

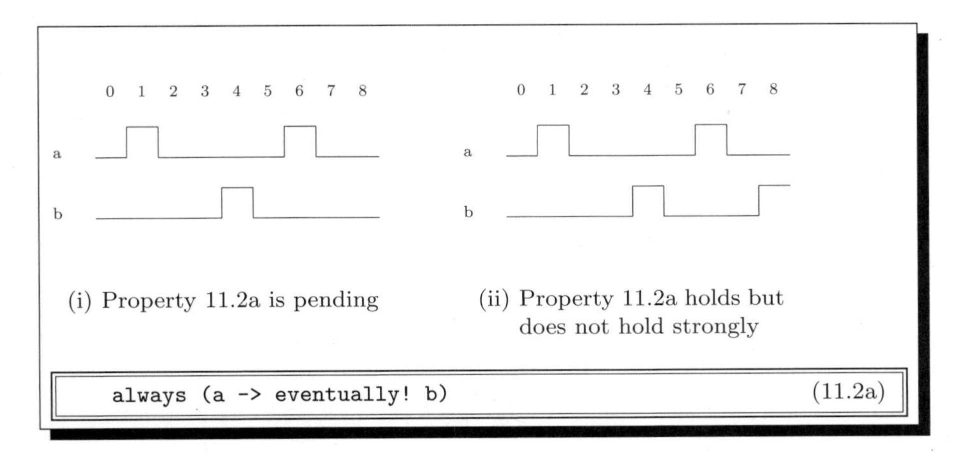

(i) Property 11.2a is pending

(ii) Property 11.2a holds but does not hold strongly

```
always (a -> eventually! b)                                  (11.2a)
```

Fig. 11.2: A pending property

A property whose outermost operator is `eventually!` never fails, because extending the trace may reveal a cycle on which the required sub-property holds. Conversely, a property whose outermost operator is `always` cannot hold strongly, because we can never guarantee that extending the trace will not undo the holding of the property. Since most properties used in practice have the `always` operator as their outermost operator, the status of most properties on a finite trace will be either holds but does not hold strongly, pending, or fails.

11.2 LTL style vs. SERE style

Many properties can be expressed in both LTL and SERE style with equal ease. Table 11.1 shows some common equivalences between LTL and SERE style, as well as two mixed-style properties and their equivalent in SERE style.

Sometimes we are asked the question: "Can every property containing SEREs be translated into LTL style and vice versa?". The short answer is that while every LTL style property can be translated into SERE style, not every SERE style property can be translated into LTL style. The reason is that properties which involve some form of counting, such as "signal `a` is asserted on every even cycle" cannot be expressed in PSL without using SEREs. This may seem surprising, as a beginning user might try Assertion 11.3a. Assertion 11.3a says that signal `a` must be asserted at cycle 0 (the left-hand side of the `&&`), and that whenever signal `a` is asserted, it must be asserted two cycles following (the right-hand side of the `&&`). However, that is too much. For instance, Assertion 11.3a does not hold on Trace 11.3(i), even though signal `a` is asserted on every even cycle of it. The reason is that in Trace 11.3(i), signal `a` is asserted not only at every even cycle, but also at cycle 5 (our English

Table 11.1: Some common equivalences between LTL and SERE style. In the table, `b` and `c` designate Boolean expressions, `p` designates a property, `s` designates a SERE, and `i` and `j` designate integers.

LTL style property	SERE style equivalent
`eventually! b`	`{[*] ; b}!`
`eventually! s!`	`{[*] ; s}!`
`b until c`	`{b[*] ; c}`
`b until s`	`{b[*] ; s}`
`b until! c`	`{b[*] ; c}!`
`b until! s!`	`{b[*] ; s}!`
`b && next c`	`{b;c}`
`b && next s`	`{b;s}`
`b && next! c`	`{b;c}!`
`b && next! s!`	`{b;s}!`
`next[i] (b)`	`{[*i] ; b}`
`next[i] (s)`	`{[*i] ; s}`
`next![i] (b)`	`{[*i] ; b}!`
`next![i] (s!)`	`{[*i] ; s}!`
`next_a[i:j](b)`	`{[*i] ; b[*j-i+1]}`
`next_a[i:j](s)`	`for k in {i:j}: & {[*k] ; s}`
`next_a![i:j](b)`	`{[*i] ; b[*j-i+1]}!`
`next_a![i:j](s!)`	`for k in {i:j}: & {[*k] ; s}!`
`next_e[i:j](b)`	`{[*i:j] ; b}`
`next_e[i:j](s)`	`{[*i:j] ; s}`
`next_e![i:j](b)`	`{[*i:j] ; b}!`
`next_e![i:j](s!)`	`{[*i:j] ; s}!`
`always {s} |-> p`	`{[*];s} |-> p`
`always {s} |=> p`	`{[*];s} |=> p`

specification does not prohibit this). However, the right-hand side of the `&&` in Assertion 11.3a requires that if this happens, then signal `a` should then be asserted at cycle 7 as well. On the other hand, our English specification can easily be expressed in SERE style as shown in Assertion 11.3b. Note that we have interpreted even cycles as those numbered with even numbers, where numbering of cycles starts from 0. Thus Assertion 11.3b requires that `a` be asserted on the first cycle of the trace (cycle 0), on the third (cycle 2), fifth (cycle 4) and so on.

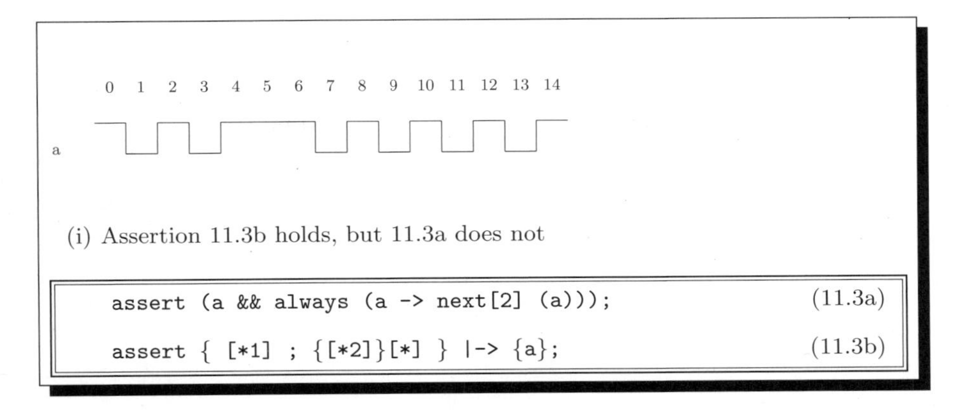

(i) Assertion 11.3b holds, but 11.3a does not

```
assert (a && always (a -> next[2] (a)));      (11.3a)

assert { [*1] ; {[*2]}[*] } |-> {a};          (11.3b)
```

Fig. 11.3: Expressing the requirement that a hold on every even cycle

```
vunit every_even {

   reg even;
   initial even <= 'true;
   always @(posedge clk)
      even <= !even;

   assert always (even -> a);

}
```

Fig. 11.4: Using modeling code to check that a holds at every even cycle

The long answer, however, is not so clear cut. First of all, while theoretically every LTL style property can be translated into SERE style, in practice it is not always so simple to do so. And while, as we have seen above, not every SERE style property can be translated into LTL style, it is possible to do so with the addition of modeling layer code. For instance, by adding modeling layer code as shown in vunit `every_even` in Figure 11.4, we can declare an auxiliary signal `even` that allows us to achieve our goal.

11.3 Common equivalences in LTL style

Saying that signal `a` is asserted at every cycle is the same as saying that it will never be deasserted. Thus, it is easy to see that Properties 11.5a and 11.5b are equivalent. If we replace signal `a` with any PSL property `p`, the equivalence is preserved. Table 11.2 shows this and other common equivalences between two LTL style properties. Each follows immediately from the definition of the

```
always a                                                     (11.5a)

never !a                                                     (11.5b)
```

Fig. 11.5: Two equivalent properties

relevant operators. Note that in many cases, and depending in some cases on the particular properties `p` and `q`, only one of the equivalent properties shown belongs to the simple subset. This illustrates that, as stated previously, the simple subset is mostly a matter of form, not of content. Working out the equivalences shown is left as an exercise for the reader.

The table shows several weak-strong pairs of operators that are dual to one another. This is a general rule in PSL – negating a property whose outermost temporal operator is strong gives a property whose outermost temporal operator is weak, and vice versa. As before, working out the equivalences is left as an exercise for the reader.

11.4 PSL for formal verification

Most PSL features are intended for both documentation as well as verification of any kind. In this chapter, we discuss the few features that are particularly oriented for use in formal verification.

11.4.1 Infinite traces

In the rest of this book, we have been concerned with finite traces as seen in simulation or other bounded verification methods, and have ignored the issue of infinite traces as seen in formal verification. Thus, we have explained the difference between weak and strong temporal operators as coming into play when the trace is "too short" to completely determine the status of a property. Obviously, an infinite trace cannot be "too short".

On an infinite trace, a strong temporal operator can be understood in the same way it is understood on a finite trace – everything that needs to occur must occur. For instance, in order for `eventually! a` to hold on an infinite trace, we must see at least one assertion of `a` at some cycle. A weak temporal operator can be understood as signifying that the property holds on every finite prefix of the infinite trace in question. For example, `a until b` is usually understood as holding on an infinite trace if either `a until! b` or `always a` holds, which happens if and only if `a until b` holds on every finite prefix of the infinite trace.

Table 11.2: Some common equivalences. In the table, b designates a Boolean expression, p and q designate properties.

<table>
<tr><th>Property</th><th>An equivalent property</th></tr>
<tr><td>p || q</td><td>!(!p && !q)</td></tr>
<tr><td>p && q</td><td>!(!p || !q)</td></tr>
<tr><td>p -> q</td><td>!p || q</td></tr>
<tr><td>always p</td><td>!(eventually! (!p))</td></tr>
<tr><td>always p</td><td>never (!p)</td></tr>
<tr><td>eventually! p</td><td>!(always (!p))</td></tr>
<tr><td>eventually! p</td><td>`true until! p</td></tr>
<tr><td>next p</td><td>!(next! (!p))</td></tr>
<tr><td>next! p</td><td>!(next (!p))</td></tr>
<tr><td>p until q</td><td>(p until! q) || (always p)</td></tr>
<tr><td>p until q</td><td>!((!q) until! (!p && !q))</td></tr>
<tr><td>p until! q</td><td>!((!q) until (!p && !q))</td></tr>
<tr><td>p until_ q</td><td>p until (p && q)</td></tr>
<tr><td>p until!_ q</td><td>p until! (p && q)</td></tr>
<tr><td>p before q</td><td>(!q) until (p && !q)</td></tr>
<tr><td>p before! q</td><td>(!q) until! (p && !q)</td></tr>
<tr><td>p before_ q</td><td>(!q) until p</td></tr>
<tr><td>p before!_ q</td><td>(!q) until! p</td></tr>
<tr><td>next_event(b)(q)</td><td>(!b) until (b && q)</td></tr>
<tr><td>next_event!(b)(q)</td><td>(!b) until! (b && q)</td></tr>
</table>

11.4.2 The Optional Branching Extension (OBE)

We have seen SERE style and LTL style PSL properties, together known as the Foundation Language, or FL. PSL also contains an Optional Branching Extension (OBE), which includes properties from the branching temporal logic CTL. The OBE is useful only in formal verification, as it allows you to express properties such as "there exists a trace such that ..." that cannot be checked by a single simulation trace. Since the OBE is not particularly useful unless you are using a CTL model checker, we will not go into more detail here. If you are using a CTL model checker, please see any standard reference on CTL.

11.4.3 The `union` operator and the built-in function `nondet()`

The `union` operator provides a nondeterministic choice between two values. For instance, assume that signal `req` is an input to our design, and that it can have any possible behavior with the exception that it cannot be asserted for two consecutive cycles. We could specify this using an `assume` directive, as shown in vunit `req_by_assume` in Figure 11.6.

```
vunit req_by_assume {
   reg req;

   assume never {req;req};
}
```

Fig. 11.6: A vunit giving signal `req` behavior using an `assume` directive

If we would rather use a direct style of specifying the behavior of the environment, we can use the `union` operator as shown in vunit `req_by_union` in Figure 11.7. The vunit `req_by_union` assigns the value `'false` to signal `req` the cycle following an assertion of `req` and a nondeterministic value otherwise.

```
vunit req_by_union {
   reg req;

   always @(posedge clk)
      req <= req ? 'false : ('false union 'true);
}
```

Fig. 11.7: A vunit giving signal `req` behavior using the `union` operator

A nondeterministic assignment is similar in some ways to a random choice, and some tools may choose to interpret them as identical. However, for some tools, such as formal verification tools, there is an important difference. A random choice implies that one out of many possible values is chosen. A nondeterministic choice implies that all values must be taken into consideration. This can be understood as simultaneously choosing all possible values, but on different threads of execution.

NOTE: Just as the `forall` operator does not imply that replication is actually taking place in the verification tool, use of the `union` operator does not imply that threads are actually being created. Rather, the description above is intended to convey the functionality of the `union` operator, not its implementation. In fact, both formal as well as simulation-based verification tools may and frequently do implement nondeterministic choice without resorting to the creation of separate execution threads.

While the `union` operator gives nondeterministic choice between two values, the `nondet()` built-in function gives nondeterministic choice between a set of values. For instance, the call `nondet({a, b, c})` is equivalent to `(a union b union c)`.

It is also possible to create a vector of nondeterministic choices. For a number `N` and a set of values `T`, a call to `nondet_vector(N,T)` returns a vector of length `N`, whose elements are chosen nondeterministically from the set `T`. For instance, `nondet_vector(8,{'true,'false})` returns a vector of length 8, each element of which is chosen nondeterministically as either `'true` or `'false`.

While nondeterministic choice is particularly useful in formal verification, it can be useful as well in simulation. For instance, a simulation tool could choose to interpret a nondeterministic assignment to a signal as a random choice, or it could choose to interpret nondeterministic choice in a more sophisticated manner, for instance as an instruction to run multiple simulations, each of which uses one of the possible values.

11.4.4 The built-in function `next()`

The built-in function `next()` returns the value that a signal will hold at the next PSL cycle. Thus a call to `next()` with parameter `a` (`next(a)`) will return the value that signal `a` will hold the following cycle. If signal `a` is deterministic, then the value of `next(a)` will typically be available in another way. For instance, if `a` is the output of a flip-flop, then `next(a)` can be used interchangeably with the input of the flip-flop.

The utility of `next()` is thus not to access the next value of deterministic signals like design signals, but rather to access the next value of signals that are nondeterministic – for instance, signals defined in the modeling layer. For example, suppose that signal `b` is a nondeterministic signal, and suppose that signal `c` is a signal that should shadow the value of signal `b` in the following way: it should have the value `'false` the cycle following an assertion of `ignore_b`, and the value of `b` otherwise. We can code this in PSL as shown in vunit `good_shadowing` of Figure 11.8.

Note that we cannot achieve the desired linkage without `next()`. For instance, if we try coding an assignment to `c` as shown in vunit `bad_shadowing` in Figure 11.9, we do not get what we want. The reason is that in the case that both `ignore_b` and `en` are deasserted, signal `c` does not necessarily shadow signal `b`: because of the two independent calls to `nondet()`, signals `b` and `c` could end up with different values in such a case.

While built-in function `next()` is particularly useful in formal verification, it could also be used in simulation environments as a way to link two signals one of which has a random value, or even one of which has a nondeterministic

```
vunit good_shadowing {
   reg b, c;

   always @(posedge clk)
      if (en)
         b <= new_b;
      else
         b <= nondet({'false,'true});

   always @(posedge clk)
      if (ignore_b)
         c <= 'false;
      else
         c <= next(b);
}
```

Fig. 11.8: An illustration of how to shadow the value of `b` using `next()`

```
vunit bad_shadowing {
   reg b, c;

   always @(posedge clk)
      if (en)
         b <= new_b;
      else
         b <= nondet({'false,'true});

   always @(posedge clk)
      if (ignore_b)
         c <= 'false;
      else if (en)
         c <= new_b;
      else
         c <= nondet({'false,'true});
}
```

Fig. 11.9: An illustration of how NOT to shadow the value of `b`

value, in the case that nondeterministic values are used in simulation (see Section 11.4.3 above).

11.4.5 Vunit scoping rules

Consider the vunit `illustrate_scoping_rules` shown in Figure 11.10. The vunit `illustrate_scoping_rules` is bound to `block1` of module `top`. If signal `b` is a signal of `block1`, the assignment appearing in the vunit over-

```
vunit illustrate_scoping_rules(top.block1) {
   wire b;
   assign b = 'false;

   assert ...
}
```

Fig. 11.10: An illustration of the utility of the scoping rules for formal verification

rides it. That is, in a verification run of any assertions appearing in vunit `illustrate_scoping_rules`, the true value of signal `b` in the design will be ignored, and the value given in the vunit (in this case, `'false`) will be used. Providing a mechanism to override design signals in this way allows easy masking out of part of the design without the need to change the RTL itself. This is useful, for instance, when size problems prevent the formal verification of a whole block or unit as one piece.

While overriding design signals is particularly useful in formal verification, it can be useful in simulation as well. For instance, if test generation is being done from an environment described in modeling code, then overriding design signals can be used to create tests that focus only on a piece of a whole block or unit, without the need to change the hierarchical boundaries of the RTL code.

11.4.6 Safety and liveness

Intuitively, a *safety property* is used to ensure that "something bad doesn't happen", while a *liveness property* is used to ensure that "something good eventually happens". Using only non-negated weak operators results in a safety property, and using non-negated strong operators usually results in a liveness property. Some properties are neither safety nor liveness – they are *mixed* properties. For instance, the property `always (a -> eventually! b)` is a mixed property. The concepts of safety and liveness are important in formal verification because safety properties are easier to verify than liveness or mixed properties.

11.5 Vacuity

Vacuity is not a concept peculiar to PSL, and it is not defined as part of the language. However, because it is a useful concept to users of assertions in general, and because many tools that support PSL also support vacuity, we briefly touch on it here.

A *vacuous pass* of an assertion is an assertion that holds, but in a trivial manner. For instance, consider Property 11.11a. Property 11.11a states that

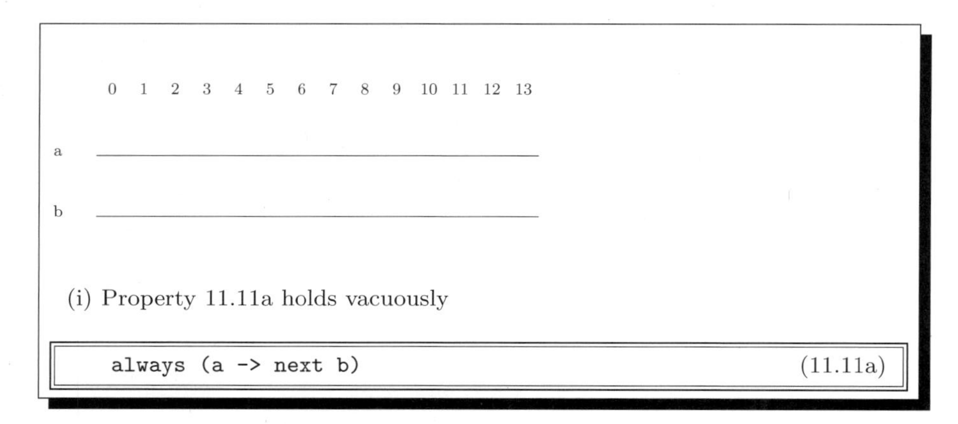

Fig. 11.11: Vacuity

whenever `a` holds, `b` will hold on the following cycle. Recall that a logical implication defaults to true (that is, `a -> next b` holds in a cycle if `a` does not hold in that cycle). Thus, Property 11.11a holds on Trace 11.11(i): there are no assertions of `a`, therefore the sub-property `a -> next b` holds in all cycles, and thus the entire property, which applies `always` to `a -> next b`, holds.

However, there is something very fishy about the holding of Property 11.11a on Trace 11.11(i). Surely someone who went to the trouble of coding Property 11.11a intended that there be at least one assertion of `a` at some cycle. The pass of Property 11.11a on Trace 11.11(i) is a vacuous pass because signal `a` was never asserted, thus allowing the property to hold in a trivial manner.

If the left-hand side of a suffix implication never holds, it can also cause a vacuous pass. For instance, Property 11.12a holds vacuously on Trace 11.12(i) because the SERE {`a;b;c`} never happens, and thus the right-hand side {`d`} is never required to hold.

In general, vacuity can be understood as follows: if a passing property contains a Boolean expression that has no effect on whether or not the property holds (it could be switched with any other Boolean expression, including `'false` and `'true`, without causing the property to fail), then the pass is vacuous. For example, the `d` of Property 11.12a could have been switched to `'false`, and the resulting property `always {a;b;c} |=> {'false}` would still hold on Trace 11.12(i).

As another example, consider Property 11.13a. Property 11.13a has two implications, whose left-hand sides are `req` and `ack`. Both left-hand sides hold at some point in Trace 11.13(i), however, the timing is such that the sub-property `eventually! gnt` is never required to hold. Thus, Property 11.13a holds vacuously on Trace 11.13(i).

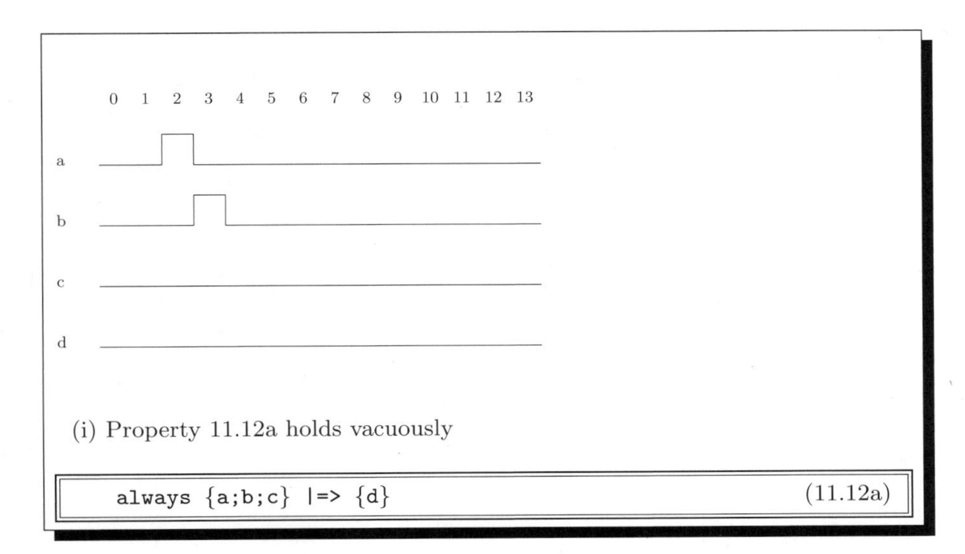


```
always {a;b;c} |=> {d}
```
(11.12a)

Fig. 11.12: A vacuous pass caused by the left-hand side of a suffix implication never holding

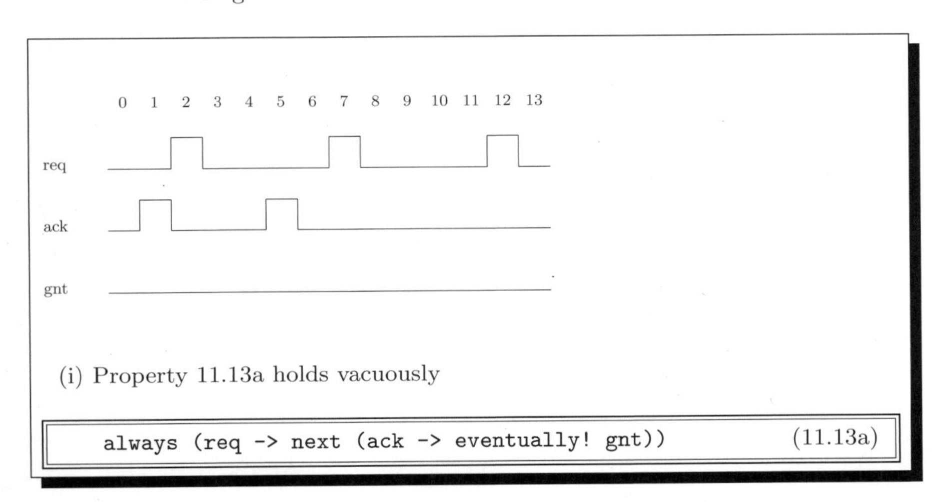


```
always (req -> next (ack -> eventually! gnt))
```
(11.13a)

Fig. 11.13: A vacuous pass caused by the timing between two left-hand sides not lining up

Up until now, all of our examples have included some form of implication. Vacuity is more than the left-hand side of a logical implication or suffix implication never holding, however. For example, consider the property `never {req;req}` on a trace where `req` never holds. In such a case, the second `req` is superfluous, and thus `never {req;req}` holds vacuously on such a trace.

Notice that a property that passes vacuously on one trace may pass non-vacuously on another. Thus, depending on the design, the property, and the intention of the person who coded the property, vacuity is probably of interest only if all of the passes of a particular property are vacuous. In such a case, the vacuity may indicate a problem in the design (such that some set of expected sequence of events is not being seen), a problem in the test suite or driving environment of the verification (such that some input sequences are not being generated correctly), or a problem of coverage (such that some scenarios are not being generated at all).

11.6 Flavor issues

While mixing flavors in a PSL specification is not allowed (otherwise the specification could not be parsed), there is nothing that prohibits the application of a PSL specification written in flavor A to a design written in flavor B. Indeed, PSL specifically allows this. When flavors are mixed in this way, the standard conventions for mixed language simulation are used when a Boolean expression in flavor A refers to a signal from a design written in flavor B.

12

More Philosophy – High- vs. Low-level Assertions

So now you know PSL. What are you going to do with it? In this chapter, we discuss the what, rather than the how, of specifying in PSL using three simple but real-world examples.

12.1 A simple state machine

Suppose that the state machine shown in Figure 12.1 is part of your specification, and suppose that you are interested in writing PSL assertions to check the implementation. A common way to proceed would be to write an assertion about every transition in the state machine, like those shown in Figure 12.2. Such assertions are very low-level: they check that a few gates of the design function according to a few lines of the specification.

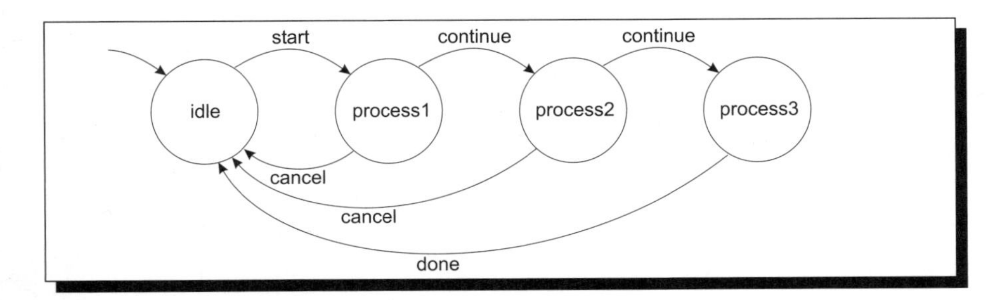

Fig. 12.1: A state machine for processing commands

While low-level assertions can be useful, writing higher level assertions can be much more powerful. For instance, the low-level assertions do not ensure that processing ever completes. In all likelihood, that is an unspoken assumption of our specification. From Figure 12.1, we see that the signals

```
assert always ((state=='idle && start) ->                (12.2a)
      next (state=='process1));

assert always ((state=='idle && !start) ->               (12.2b)
      next (state=='idle));

assert always ((state=='process1 && continue) ->         (12.2c)
      next (state=='process2));

assert always ((state=='process1 && cancel) ->           (12.2d)
      next (state=='idle));

assert always                                            (12.2e)
      ((state=='process1 && !(continue || cancel)) ->
      next (state=='process1));

  .
  .
  .
```

Fig. 12.2: A property about each transition

controlling whether or not processing completes are `continue`, `cancel`, and `done`. We could write assertions about these signals and their relations to the states of our FSM, for instance, that if we are in state `process1` or in state `process2` we eventually get either a `continue` or a `cancel`, and if we are in state `process3` we eventually get a `done`. However, there is a much simpler way. We can simply write a single assertion that states that if we are not in state `idle`, we will eventually get back to it, as shown in Assertion 12.3a. Assertion 12.3a is a high-level assertion. Whether it holds or not is not the responsibility of a few lines of code, but rather of the whole system. If it fails, the problem could be in the state machine or in the code that generates the control signals `continue`, `cancel` and `done`, or perhaps somewhere else – for instance in the code that generates the signals driving the code that generates the control signals. Assertion 12.3a is a simple and rather obvious assertion that usually goes unsaid in English language specifications. However, as a PSL assertion it can be very useful in flushing out bugs, because it does not focus

```
assert always ((state!='idle) ->                    (12.3a)
     eventually! (state=='idle));
```

Fig. 12.3: Making sure that processing always completes

on a few lines of code or even on a particular block or unit, but instead takes a high-level view of the entire chip or system under verification.

We've written a low-level assertion about each transition, and in addition a high-level assertion that ensures that processing eventually completes. Should we be satisfied? The answer is no. Every piece of code has a purpose. The purpose of our FSM is to aid in the processing of commands. Just because the state machine steps through the expected states and eventually completes does not mean that the commands are being processed correctly. There could be a problem in the design of the FSM, even if its implementation matches Figure 12.1.

For example, suppose that we have a command type called `read`, and suppose that after a read is completed, we expect to have the read data in memory. Correct processing of the read might be affected by a bug in buffer allocation, or a bug in which our state machine is never triggered – i.e., a `start` is never issued for it. Does this mean that we should write assertions ensuring that buffers are allocated as per the specification, and that for every read, signal `start` is asserted?

Maybe. But such assertions are again very low-level. If they hold, we will know that the designer has faithfully implemented what is written in the specification. However, we will not know whether or not what is written in the specification actually achieves the desired functionality. As an extreme and obviously absurd example, let's say that the design is intended to be a pipelined multiprocessor, however by mistake the architects have architected a pocket calculator. In such a case, all low-level assertions may pass, but of course the desired functionality has not been achieved. Writing high-level assertions ensures that a more subtle hole in the specification will not be missed.

Thus, to return to our example, we should augment our low-level assertions with high-level assertions such as: if a read is issued and the current value of the data is `X`, then eventually the read will complete and the data that is returned will be `X`. Like Assertion 12.3a, such an assertion frequently goes unsaid in an English language specification, but is a very powerful PSL assertion. While checking what the specification says is very important, checking what it does not say is probably more so.

Notice that if the high-level assertions hold, the low-level assertions could be redundant: if the desired functionality is achieved, but the state machine does not exactly match that of the specification, we might not care. This does not mean, however, that low-level assertions shouldn't be written. Low-level assertions have their place, since they are usually easy to write (they follow directly from the text of the specification) and the problems they uncover are usually easy to debug (the problem can be pinpointed to a particular line in the code – the one, for instance, where the desired transition failed to take place). However, because low-level assertions do not check that the desired functionality has been achieved, they are not enough – you should always be sure to include the appropriate high-level assertions as well.

12.2 A simple FIFO

As another example, consider a FIFO. A beginner might be tempted to write assertions that check exactly what happens to the FIFO entries when a load or an unload is received. Such assertions are low-level, depend upon the particular implementation, and do not capture the essence of the FIFO – its point. A better way would be to count the number of entries that should be valid using the modeling code, then write a single assertion that says that if there are `i` valid entries in the FIFO and a load is received, then the $i+1^{th}$ unload should return the value loaded. For instance, assuming that a load (assertion of signal `load`) should load the value in `data_in[63:0]` and that an unload (assertion of signal `unload`) should unload a value into `data_out[63:0]`, and that furthermore our FIFO allows immediate unloading of a value loaded in the same cycle (in the case the FIFO is empty), we can code a `vunit` as shown in Figure 12.4. We declare an auxiliary signal `count[3:0]` in the modeling layer that keeps track of how many items are in the FIFO. Signal `count[3:0]` is incremented if `load` is asserted and `unload` is not, it is decremented if `unload` is asserted and `load` is not, and it keeps its value otherwise. Note that we have taken into consideration the case that `load` and `unload` are asserted together – in this case the count should not change.

The first two assertions in vunit `high_level_fifo_check` of Figure 12.4 check that signal `count[3:0]` does not over- or underflow. This might happen because of a bug in the design, but it also might happen because of a bug in the modeling code – perhaps our FIFO is more than 15 entries deep, so we should have declared `count` to be wider. Note that no over- or underflow can occur as a result of `load` and `unload` being asserted together, thus we have coded our assertions in such a way that they cannot fail in this case.

The last assertion in vunit `high_level_fifo_check` is our high-level assertion. It checks that whenever we have `i` items in the FIFO, and a value of `DAT[63:0]` is loaded, then the $i+1^{th}$ assertion of `unload` should unload the value `DAT[63:0]`. For example, if a load occurs and the FIFO is empty, then the next unload should return the data just loaded. If a load occurs and the FIFO contains one item, then the second unload should return the data just loaded, and so on. Recall that the `next_event` operator includes the current

```
vunit high_level_fifo_check {
   reg [3:0] count;
   initial
      count[3:0] <= 4'b0;

   always @(posedge clk)
      if (load & !unload)
        count[3:0] <= count[3:0] + 1;
      else if (!load & unload)
        count[3:0] <= count[3:0] - 1;

   assert always ((count[3:0]==0) -> !(!load & unload));
   assert always ((count[3:0]==15) -> !(load & !unload));

   assert forall i in {0:15}:  forall DAT[63:0] in boolean:
      always
         ((count[3:0]==i && load && data_in[63:0]==DAT[63:0])
         ->
         next_event(unload)[i+1](data_out[63:0]==DAT[63:0]));
}
```

Fig. 12.4: A vunit to check the high-level functionality of a FIFO

cycle. Thus, if an unload occurs the same cycle as the load, it will be counted as one of the `i+1` unloads.

Recall from Section 8.3 that a replicated property like that used in the third assertion of vunit `high_level_fifo_check` of Figure 12.4 does not necessarily imply that any replication is actually taking place. This is extremely important to understand in the case of real-life examples like that of Figure 12.4. If replication was taking place, it would imply that a tool would need to generate 16×2^{64} separate properties. Obviously, this would not work in practice. If your tool implements replicated properties by actual replication and therefore cannot handle properties like that of the third assertion in vunit `high_level_fifo_check`, be sure to complain.

Besides efficient implementations of replicated properties, the vunit of Figure 12.4 uses another feature of PSL that may not be supported by all tools: the use of the forall parameter `i` in combination with the `next_event` operator to designate the $i + 1^{th}$ next event. In case your tool does not support this, you can use the `%for` macro to create 16 separate assertions, as shown in Figure 12.5. The vunit `workaround` of Figure 12.5 is functionally equivalent to vunit `fifo` of Figure 12.4. However, vunit `workaround` uses `%for` to explicitly create 16 different properties. Notice that we used `%for` instead of `forall` for `i`, but not for `DAT[63:0]`, thus we have created 16 separate properties, but not 16×2^{64}, since use of the `forall` operator does not require replication.

```
vunit workaround {
   reg [3:0] count;
   initial
      count[3:0] <= 4'b0;

   always @(posedge clk)
      if (load & !unload)
        count[3:0] <= count[3:0] + 1;
      else if (!load & unload)
        count[3:0] <= count[3:0] - 1;

   assert always ((count[3:0]==0) -> !(!load & unload));
   assert always ((count[3:0]==15) -> !(load & !unload));

%for i in 0 .. 15 do
   assert forall DAT[63:0] in boolean:
      always
         ((count[3:0]==i && load && data_in[63:0]==DAT[63:0])
         ->
         next_event(unload)[i+1](data_out[63:0]==DAT[63:0]));
%end
}
```

Fig. 12.5: A vunit with the same functionality as that of Figure 12.4, but using `%for i` instead of `forall i`

12.3 A simple bus interface

As a final example of high- vs. low-level assertions, consider the following simple bus interface: there are two commands – read and write. A read is indicated by assertion of signal `read`, and a write is indicated by assertion of signal `write`. A command can be issued only after requesting the bus, indicated by a pulsed assertion of signal `bus_req`, and receiving a grant, indicated by the assertion of signal `gnt` one cycle after the assertion of `bus_req`. If the bus was not requested, it shouldn't be granted. The command is issued the cycle following receipt of the grant. Either a read or a write can be issued, but not both simultaneously. Reads and writes come with an address, on `addr[7:0]`, that should be valid the following cycle. Address validity is indicated by signal `addr_valid`. If a read is issued, then one pulse of data on `data_in[63:0]` is expected the following cycle, and if a write is issued, then one pulse of data on `data_out[63:0]` is expected the following cycle. Valid read data is indicated by `data_in_valid`, and valid write data by `data_out_valid`. We can code up this specification as shown in vunit `bus_check` of Figure 12.6, which contains low-level assertions that verify that our implementation follows the bus protocol. However, what about the purpose of the bus? It is conceivable that our bus follows the protocol perfectly, but our chip does not work as expected,

```
vunit bus_check {
   assert always ((read || write) -> ended({bus_req;gnt;'true}));
   assert always (!bus_req -> next !gnt);
   assert never (read && write);
   assert always ((read || write) -> next (addr_valid));
   assert always (!(read || write) -> next (!addr_valid));
   assert always (read -> next (data_in_valid));
   assert always (!read -> next (!data_in_valid));
   assert always (write -> next (data_out_valid));
   assert always (!write -> next (!data_out_valid));
}
```

Fig. 12.6: A vunit to check the protocol of our simple bus

because reads and writes are not issued when needed, or because they are abandoned when a grant is not issued, or because bad data is returned.

In order to make sure that we not only follow the protocol, but also achieve the desired functionality, we need to write high-level assertions about the purpose of the design. Presumably, we can tell when a read or a write is necessary. For this simple example, let's assume that we have two input signals `get_data` and `put_data`, that indicate that a read or a write, respectively, are needed. Furthermore, assume that we have a way to recognize, at the assertion of `get_data` and `put_data`, the data that needs to be read or written. For instance, we might know that the data that needs to be read is the data resident in memory at the address indicated by signal `get_addr[7:0]` and that it should be read into read buffer `read_buffer[63:0]`, and that the data that needs to be written is the data on signal `write_buffer[63:0]`, and that it should be written to the memory at address `put_addr[7:0]`. Furthermore, assume that array `reg [63:0] mem[0:255]` is the memory.

We could proceed by writing a bunch of low-level assertions that check that when a read is issued, the address matches that on `get_addr[7:0]`, and similarly for a write, that if a bus request is not granted, it is reissued, and so on. However, we should also code some high-level assertions that check the functionality. For instance, the two assertions of vunit `high_level_bus_check` of Figure 12.7 check that if a read is needed, then eventually the data at the location needing to be read will appear on `read_buffer` (for simplicity, we have assumed that the value in the memory cannot change between being read and appearing on signal `read_buffer`), and that if a write is needed, then eventually the data needing to be written will have been recorded at the required address in memory.

As with our previous example, recall that the use of the `forall` operator does not imply that any replication of properties is actually taking place. If replication was taking place, it would imply that a tool would need to generate

```
vunit high_level_bus_check {

   assert forall ADR[7:0] in boolean:
      always ((get_data && get_addr[7:0]==ADR[7:0]) ->
         eventually! (
           read_buffer[63:0]==mem[ADR[7:0]]));

   assert
      forall ADR[7:0] in boolean:
      forall DAT[63:0] in boolean:
         always
            ((put_data &&
              put_addr[7:0]==ADR[7:0] &&
              write_buffer[63:0]==DAT[63:0]) ->
                 eventually! (mem[ADR[7:0]]==DAT[63:0]));
}
```

Fig. 12.7: A vunit to check the high-level functionality of our simple bus

256 separate properties for the first assertion, and 256×2^{64} properties for the second. Obviously, this would not work in practice.

12.4 Summing up

To summarize the point of this chapter, a high-level assertion is an assertion that does not focus on a few lines of code or on a particular block or unit, but instead takes a high-level view of the desired functionality of the chip or system. High-level assertions are important because they check things that low-level assertions cannot check – that the desired functionality is achieved. By doing so, they can also plug holes in the low-level assertions: if a low-level assertion is missing but the desired functionality is achieved, we probably don't care, and if the desired functionality is not achieved, debugging the failure will almost always uncover a missing low-level assertion. High-level assertions do not replace low-level assertions because low-level assertions can be easier to write and to debug, however, they are extremely important in order to ensure that you are making the most of the power that assertions make available to you.

13

Common Errors

In this chapter, we discuss some common errors that beginning users of PSL tend to make, and provide some guidelines for avoiding them.

13.1 Common errors with implications

As we have seen, the logical implication (`->`) and suffix implication operators (`|->` and `|=>`) are very useful for stating if-then kinds of properties in a concise manner. However, beginning users often misuse them. In this section we examine various ways in which beginning users abuse logical and suffix implications.

13.1.1 Confusing a logical implication with a suffix implication

Both logical implications and suffix implications can be thought of as if-then expressions, with the else-part defaulting to true. The difference is in the current cycle of the then-part. The current cycle of the then-part of a logical implication is the same as the current cycle of its if-part. The current cycle of the then-part of a suffix implication is the first cycle of the *suffix* of the trace that remains once the if-part has been seen. Suppose, for instance, that we want to state that once we see a request (assertion of `req`) followed by an acknowledge (assertion of `ack`) followed by a grant (assertion of `gnt`), we expect to see a complete data transfer (assertion of `data` for four cycles followed by an assertion of `dataend`). A beginner might code Assertion 13.1a. This is a mistake! The reason is that Assertion 13.1a uses a logical implication, and therefore requires that the data transfer start the same cycle that `req` is asserted. Thus, Assertion 13.1a holds on Trace 13.1(i), but not on Trace 13.1(ii). If we mean the situation in Trace 13.1(ii), we need to use the non-overlapping suffix implication operator, as shown in Assertion 13.1b. If we meant for the

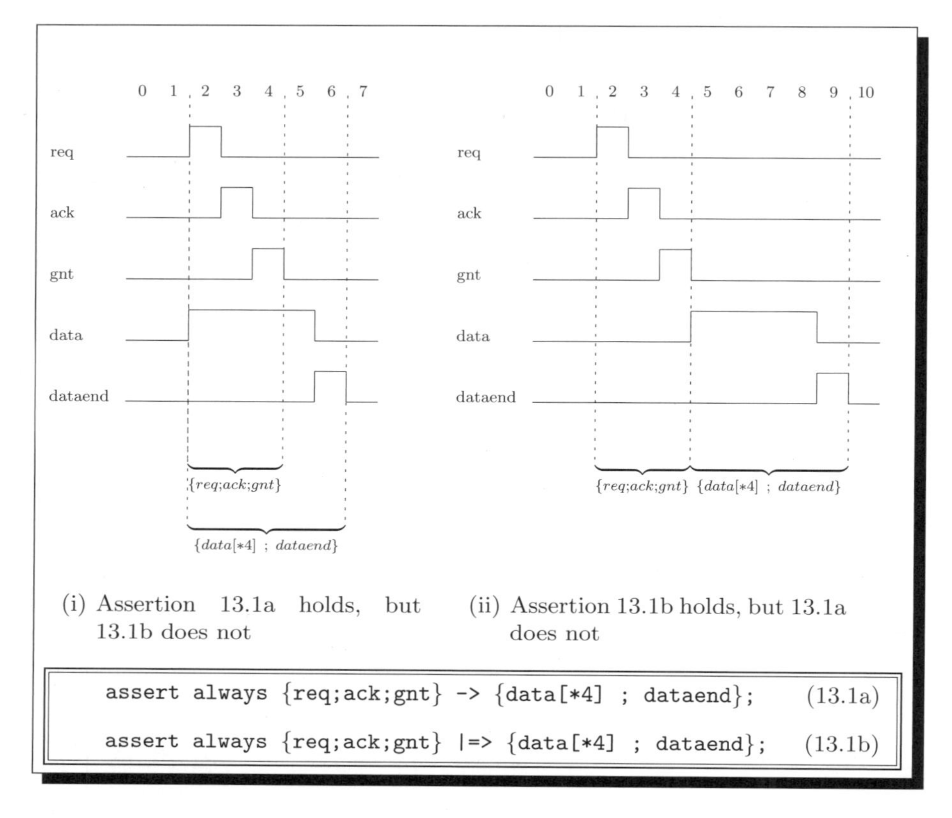

Fig. 13.1: Confusing a logical implication with a suffix implication

data transfer to start the same cycle that `gnt` is asserted rather than the cycle *after* `gnt` is asserted, we could of course have used the overlapping suffix implication operator (`|->`).

TIP: Don't use the implication operator when you mean suffix implication, and vice versa.

13.1.2 Confusing a logical implication with an "and"

Beginning users sometimes use the implication operator when they mean to use "and". Consider for instance the specification "every high priority request (assertion of `req` together with `high_pri`) should be followed immediately by an acknowledge (assertion of `ack`) and then by a grant (assertion of `gnt`)". A beginner might try Assertion 13.2a. However, this is wrong. Assertion 13.2a says that for every high priority request, *if* there is an acknowledge then there should be a grant, and therefore it holds on Trace 13.2(i). However, we wanted to say that for every high priority request, there is both an acknowledge and

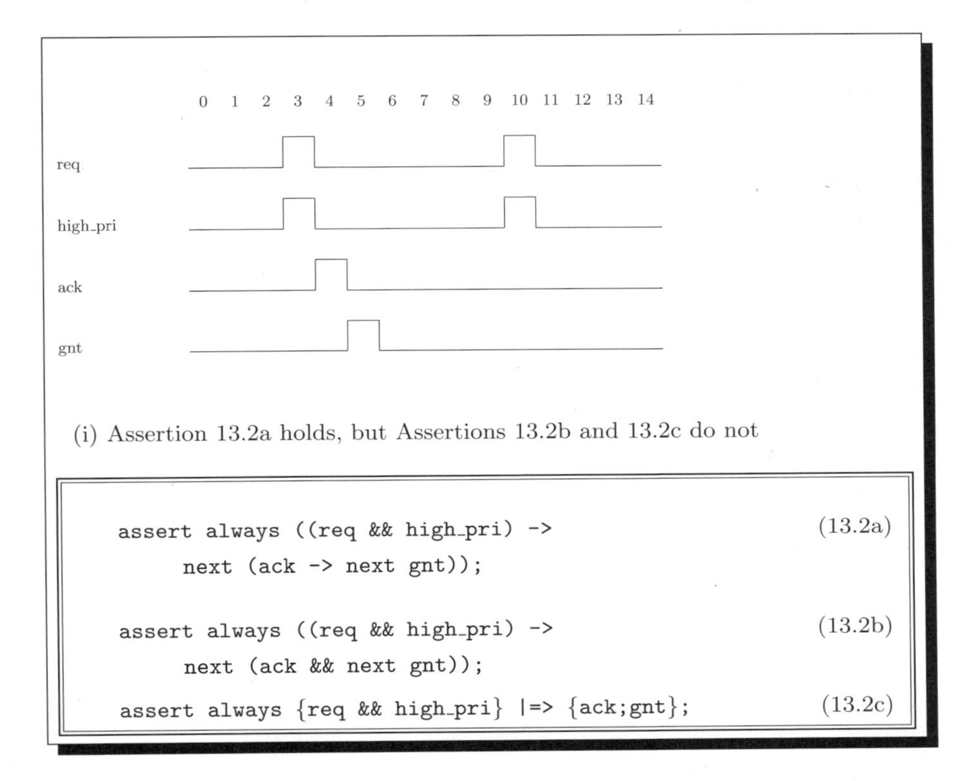

Fig. 13.2: Confusing a logical implication with an "and"

a grant. To do so, we need to use the "and" operator, as in Assertion 13.2b. Alternatively, we could have expressed the same property as shown in Assertion 13.2c. Neither Assertion 13.2b nor Assertion 13.2c holds on Trace 13.2(i) because of the missing acknowledge and grant for the second request.

TIP: Don't use a logical implication when you mean "and", and vice versa.

13.1.3 Confusing a logical implication with a concatenation

Beginning users sometimes confuse Properties 13.3a and 13.3b. The two properties are not equivalent – first, because the timing is off. The current cycle of `b` in Property 13.3a is the same as the current cycle of `c`. To see why, let's first parenthesize Property 13.3a to get the equivalent property `always (a -> ((next b) -> (next c)))` (see the operator precedence and associativity table in Appendix C). Notice that `next b` and `next c` are both sub-properties of the sub-property `(next b) -> (next c)`, connected by the Boolean `->`. Thus, `next b` and `next c` have the same current cycle. Thus, `b` and `c`, each an operand of the `next` operator, have the same current cycle, which is the

```
always a -> next b -> next c                (13.3a)

always {a;b;c}                              (13.3b)

always a -> next b -> next[2] (c)           (13.3c)
```

Fig. 13.3: Properties that are sometimes confused

cycle following the current cycle of the sub-properties `next b` and `next c`. In Property 13.3b, however, the `b` and the `c` are separated by the concatenation operator (`;`). Thus, the current cycle of `c` is the cycle following the current cycle of `b`.

Suppose we correct this problem by modifying Property 13.3a as shown in Property 13.3c. Are Properties 13.3b and 13.3c equivalent? No. We have corrected the timing, but that is not enough. Property 13.3c uses the implication operator. It says that *if* `a` holds, then *if* `b` holds the following cycle, then `c` will hold two cycles later (that is, two cycles following `a`). Property 13.3b, on the other hand, contains no implication nor suffix implication. Thus, it does not state an if-then relationship, but rather says that the sub-property `{a;b;c}` holds at every cycle. For this to happen, `a` must hold at every cycle, `b` must hold at every cycle starting from the second cycle, and `c` must hold at every cycle starting from the third cycle. As we have noted before, this is not a very useful property.

TIP: When stringing together expressions using non-SERE style, remember how the current cycle works. And never confuse a logical implication with a concatenation.

13.1.4 Confusing a suffix implication with a concatenation

Beginning users sometimes confuse the suffix implication operator with a concatenation. However, the two are very different. To see this, let's contrast Assertions 13.4a and 13.4b. Assertion 13.4a states that *if* `a` is asserted in some cycle, followed by any number of assertions of `b`, followed by an assertion of `c`, then `d` must be asserted the following cycle.

```
assert always {a ; b[+] ; c} |=> {d};       (13.4a)

assert always {a ; b[+] ; c ; d};           (13.4b)
```

Fig. 13.4: Confusing a suffix implication with a concatenation

Assertion 13.4b is very different, and as we have stated before for similar assertions, not very useful. It states that *at every cycle*, the sub-property {`a ; b[+] ; c ; d`} holds. This, in turn, requires that `a` hold at every cycle, and that following every such `a`, we see some number of assertions of `b` followed by an assertion of `c` and finally one of `d`.

TIP: Don't use a suffix implication when you mean concatenation, and vice versa.

13.1.5 Using `never` with implications

Consider Assertion 13.5a, which says that `req` is always followed by an `ack`. Suppose now that we want to state in addition that two consecutive requests are not allowed. A beginning user might think that switching the `ack` of Assertion 13.5a with `req` and the `always` of Assertion 13.5a with `never` to get Assertion 13.5b will accomplish the goal. This is a mistake! The reason is that while Assertion 13.5b indeed prohibits two consecutive assertions of `req`, it also prohibits a cycle in which `req` is deasserted.

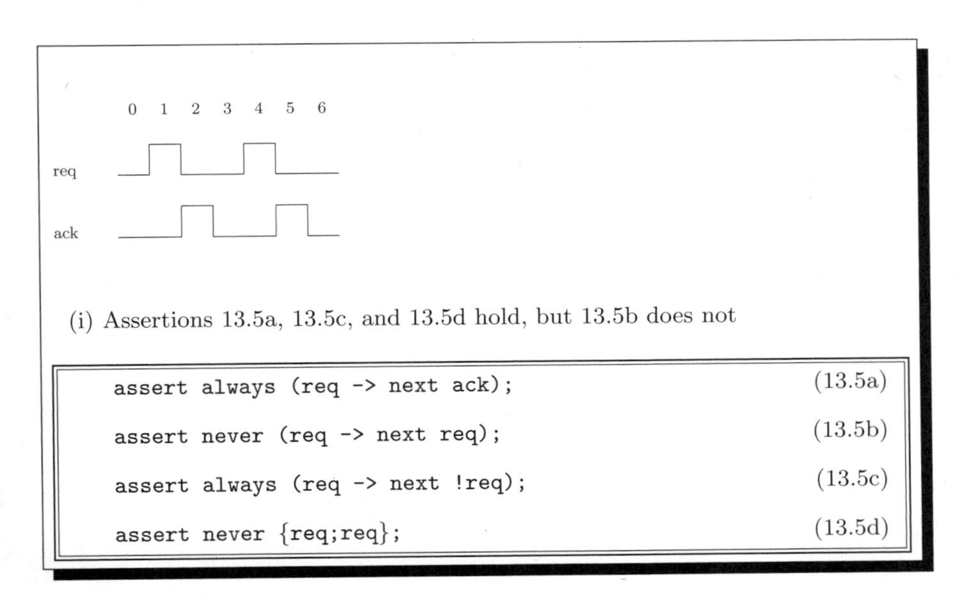

Fig. 13.5: Using `never` with logical implications

To see this, consider Trace 13.5(i). Recall that a logical implication is equivalent to an if-then expression, with the else being implicitly true. Thus, at cycle 0, the sub-property `req -> next req` holds: it is the else-part that is relevant here, and it defaults to true. In fact, the sub-property `req -> next req` holds on cycles 2, 3, 5 and 6 of Trace 13.5(i) as well. Finally, since the operand of the `never` operator holds on at least one cycle, the entire property `never (req -> next req)` does not hold at cycle 0, and thus Assertion 13.5b does not hold on Trace 13.5(i). In order to assert that there are never two consecutive requests, we can take the `never` away from the implication, and place instead a negation on the Boolean expression, as in Assertion 13.5c. Equivalently, we can use the `never` operator without applying it to a logical implication, as in Assertion 13.5d.

Consider that we want to state that a sequence of cycles matching {`a;b;c`} is always followed by a sequence of cycles matching {`d;e;f`}. We can do this in PSL as shown in Assertion 13.6a. For instance, Assertion 13.6a holds on Trace 13.6(i) because there are two sequences of cycles matching {`a;b;c`} and both of them are followed by a sequence of cycles matching {`d;e;f`}.

Suppose now that we want to state that a sequence of cycles matching {`a;b;c`} is *never* followed by a sequence of cycles matching {`d;e;f`}. A beginner might code Assertion 13.6b. However, this does not achieve our goal! To understand why, consider Trace 13.6(ii) and recall that a suffix implication

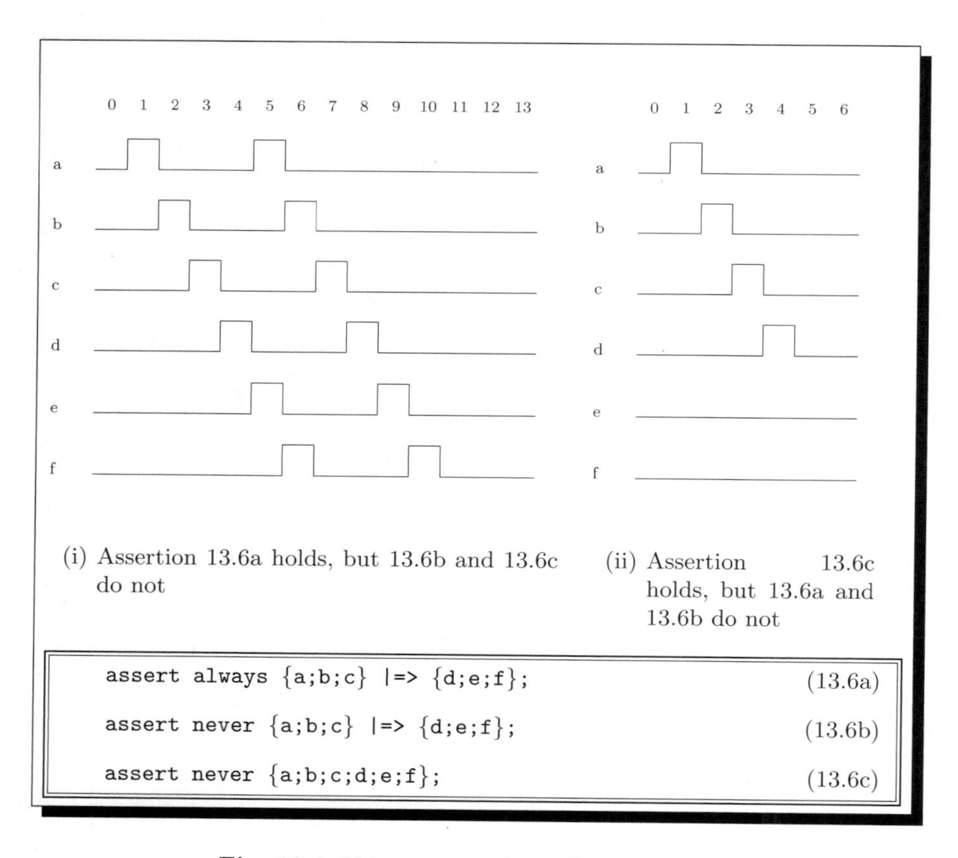

(i) Assertion 13.6a holds, but 13.6b and 13.6c do not

(ii) Assertion 13.6c holds, but 13.6a and 13.6b do not

```
assert always {a;b;c} |=> {d;e;f};     (13.6a)

assert never {a;b;c} |=> {d;e;f};      (13.6b)

assert never {a;b;c;d;e;f};            (13.6c)
```

Fig. 13.6: Using `never` with suffix implications

too can be understood as an if-then expression, with the else-part defaulting to true. Thus, at cycle 0, the sub-property `{a;b;c} |=> {d;e;f}` holds: it is the else-part that is relevant here, and it defaults to true. In fact, the sub-property `{a;b;c} |=> {d;e;f}` holds at cycles 2, 3, 4, 5 and 6 of Trace 13.6(ii) as well. Finally, since the operand of the `never` operator holds on at least one cycle, the property `never {a;b;c} |=> {d;e;f}` does not hold at cycle 0, and thus Assertion 13.6b does not hold on Trace 13.6(ii). In order to assert that a sequence of cycles matching `{a;b;c}` is never followed by a sequence of cycles matching `{d;e;f}`, we can code instead Assertion 13.6c.

Of course, it is possible to come up with some cases where applying `never` to a logical implication or to a suffix implication is exactly what you mean to say. However, such examples will usually be contrived, and can always be expressed in another manner. For these reasons, a good rule of thumb is the following:

TIP: Don't apply the `never` operator to a logical implication or to a suffix implication.

13.1.6 Negating implications

In the same way that applying `never` to a logical implication or to a suffix implication does not usually give what is intended, negating a logical implication or a suffix implication is usually a mistake.

For instance, suppose that we want to state that if we have a high priority request (assertion of signal `high_pri` together with signal `req`), then we will give an immediate acknowledge (assertion of signal `ack` the same cycle). We can state this as shown in Assertion 13.7a. For instance, Assertion 13.7a holds on Trace 13.7(i). Suppose that we now want to state that if we have a low priority request (assertion of signal `low_pri` together with assertion of signal `req`), then we will not give an immediate acknowledge. In other words, we want to write an assertion that will hold on Trace 13.7(ii). A beginner might try Assertion 13.7b. However, the beginner would be wrong. There are some cycles of Trace 13.7(ii) on which the sub-property `(low_pri && req) -> ack` holds (because at some cycles its left-hand side does not hold, and thus it defaults to true). Thus, `!(((low_pri) && req) -> ack)` does not hold at some cycles of Trace 13.7(ii), and thus Assertion 13.7b does not hold on the trace. And since the sub-property `(low_pri && req) -> ack` holds on every cycle of Trace 13.7(i), Assertion 13.7b does not hold on it either. In order to say that we will not give an immediate acknowledge, we need to place the `!` on the then-part of the implication (just as the English word "not" appears on the then-part of the English specification). Thus, Assertion 13.7c does the job and holds on Trace 13.7(ii) (as well as on Trace 13.7(i)).

In the same way that applying `!` to a logical implication is usually not a good idea, you should probably avoid the temptation of applying `!` to a suffix implication, for much the same reason. Since a suffix implication defaults to

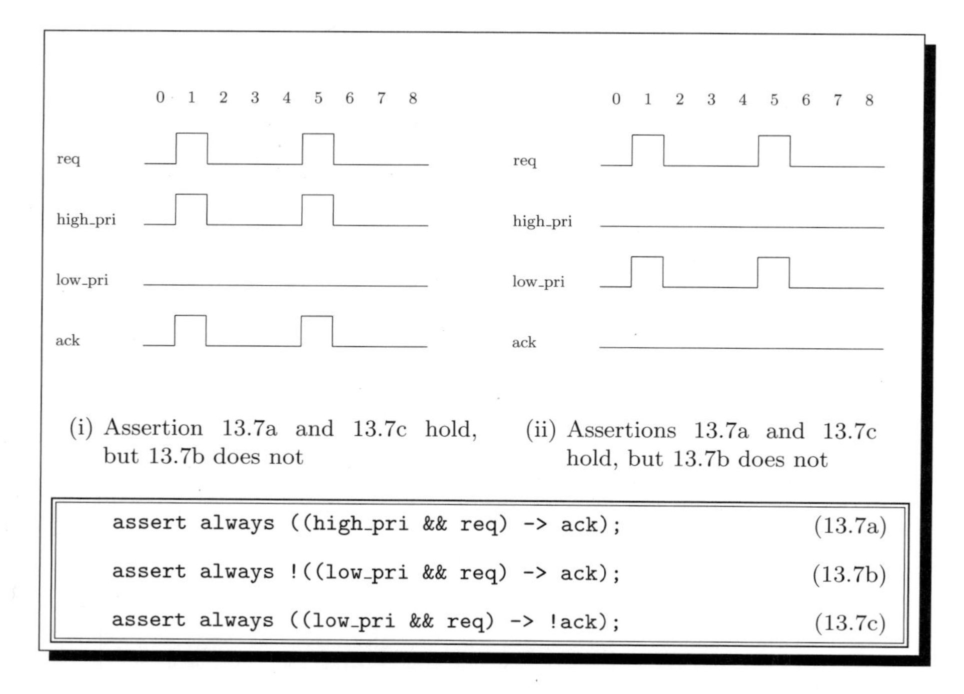

Fig. 13.7: Negating logical implications

true when its left-hand side does not hold, negating a suffix implication results in a property that "defaults to false" in such a case. As with the application of `never` to implications and suffix implications, it is possible to come up with some cases where applying `!` to a logical implication or to a suffix implication is exactly what you mean to say. However, such examples will usually be contrived, and can always be expressed in another manner. Therefore, a good rule of thumb is the following:

TIP: Don't apply `!` to a logical implication or to a suffix implication.

13.1.7 Incorrect nesting of implications

Another common error is incorrect nesting of implications and suffix implications. Consider the case that we want to say that if a request (assertion of `req`) is acknowledged (assertion of `ack` the following cycle), then it must receive a grant (assertion of `gnt` the cycle following `ack`). A beginner might try Assertion 13.8a. However, Assertion 13.8a does not hold on Trace 13.8(i), even though Trace 13.8(i) shows the scenario that we are trying to describe. Fortunately, Assertion 13.8a is not in the simple subset of PSL, thus most tools will not accept it. However, it is instructive nevertheless to understand what exactly is wrong with it aside from this.

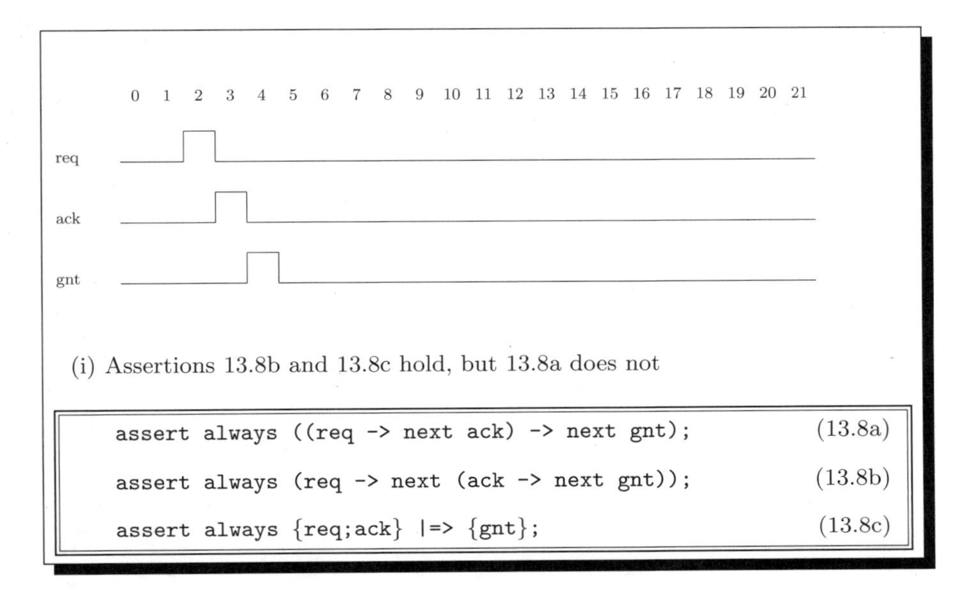

(i) Assertions 13.8b and 13.8c hold, but 13.8a does not

```
assert always ((req -> next ack) -> next gnt);        (13.8a)

assert always (req -> next (ack -> next gnt));        (13.8b)

assert always {req;ack} |=> {gnt};                    (13.8c)
```

Fig. 13.8: Incorrect nesting of implications

Assertion 13.8a does not say what we want it to say for two separate reasons. First, it sets up a logical implication between `(req -> next ack)` and `next gnt`. Thus, the current cycle of `next gnt` is the same as that of the sub-property `(req -> next ack)`, which in turn is the current cycle of `req`. In other words, whenever Assertion 13.8a requires a grant, it requires it on the cycle following a request, rather than on a cycle following an acknowledge.

The second thing wrong with Assertion 13.8a is that it requires too many grants. If there is no request on a particular cycle, `(req -> next ack)` holds (because its else-part defaults to true), and thus Assertion 13.8a will require a grant the following cycle. For example, there is no request at cycle 0 of Trace 13.8(i), therefore the sub-property `(req -> next ack)` holds at cycle 0, and thus Assertion 13.8a requires an assertion of `gnt` at cycle 1.

In order to fix Assertion 13.8a, we need to move the parentheses as in Assertion 13.8b. Assertion 13.8b says that *if* there is a request, then *if* on the following cycle there is an acknowledge, then a grant will be issued the cycle following that. It holds on Trace 13.8(i).

Assertion 13.8b illustrates the correct nesting of implications. Another way to fix Assertion 13.8a would be simply to do away with nesting of implications, and to use a suffix implication instead. Suffix implications can be easier to understand, since everything comprising the if-part appears on the left-hand side. Assertion 13.8c is equivalent to Assertion 13.8b.

Let's turn now to nesting of suffix implications. Suppose we want to state that if there is a granted read request (assertion of `read_req` followed by an assertion of `ack` followed by an assertion of `gnt`), then if there follows a complete data transfer (assertion of `data_start` followed by any number of assertions of `data` followed by an assertion of `data_end`), then the whole thing should be followed by an assertion of signal `read_complete`. A beginner might try Assertion 13.9a. However, like nesting logical implications "to the left",

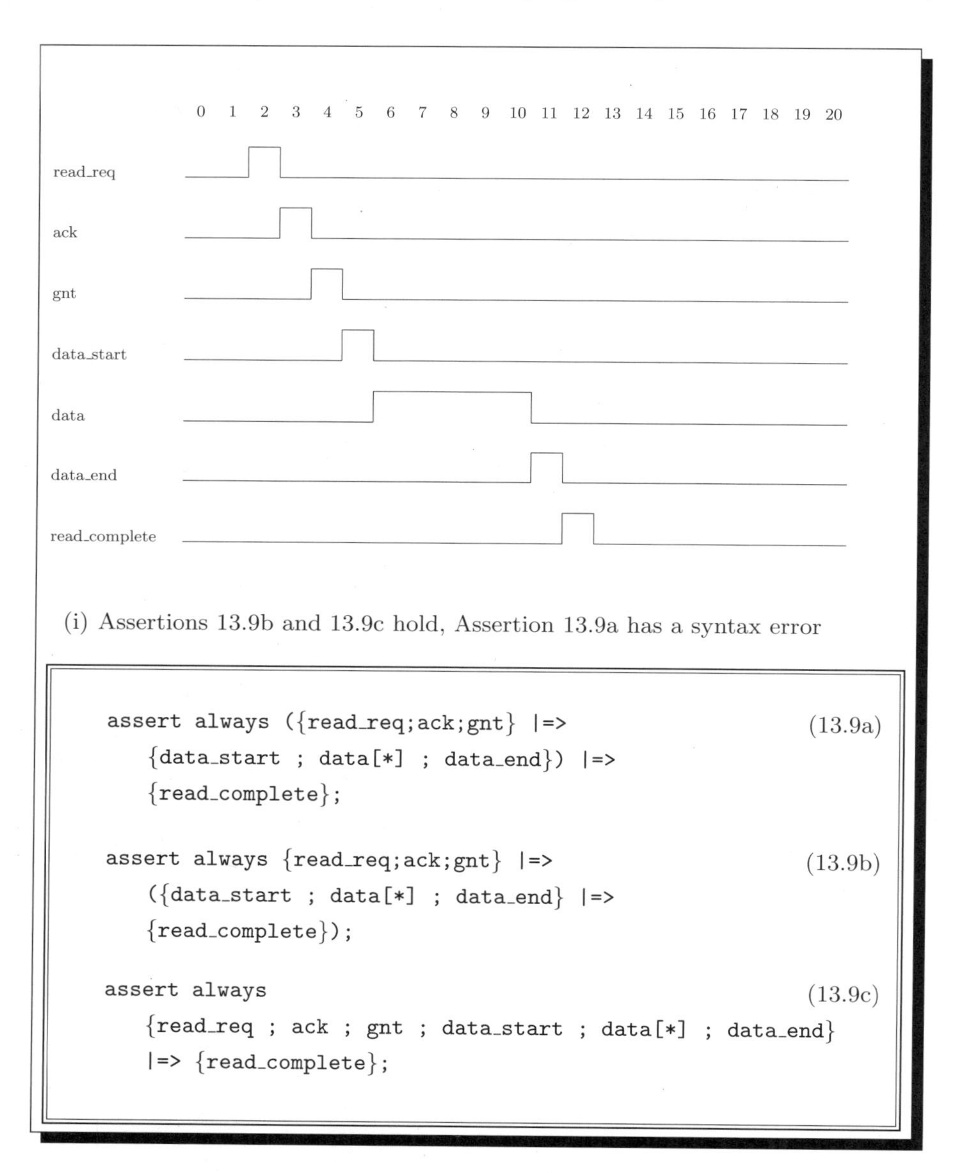

(i) Assertions 13.9b and 13.9c hold, Assertion 13.9a has a syntax error

```
assert always ({read_req;ack;gnt} |=>                          (13.9a)
   {data_start ; data[*] ; data_end}) |=>
   {read_complete};

assert always {read_req;ack;gnt} |=>                           (13.9b)
   ({data_start ; data[*] ; data_end} |=>
   {read_complete});

assert always                                                  (13.9c)
   {read_req ; ack ; gnt ; data_start ; data[*] ; data_end}
   |=> {read_complete};
```

Fig. 13.9: Incorrect nesting of suffix implications

this would be a mistake. We examined in detail why nesting logical implications in this manner does not work. For suffix implications, there is nothing to analyze – Assertion 13.9a is not in the language; it gives a syntax error because the left-hand side of the suffix implication is itself a suffix implication and not a SERE. In order to fix Assertion 13.9a, we can parenthesize "to the right" as shown in Assertion 13.9b, which says that *if* we see a read request that is granted, then *if* there follows a complete data transfer, then signal `read_complete` is asserted the cycle following the completion of the data transfer. For instance, Assertion 13.9b holds on Trace 13.9(i). There is a simpler solution, however: we can concatenate the two if-parts to get the equivalent Assertion 13.9c.

TIP: Parenthesize nested implications to the right, and avoid nesting suffix implications at all.

13.1.8 Incorrect placement of the suffix implication operator

Let's return to the previous example from a different point of view. Assertion 13.9c, repeated as Assertion 13.10a, does not require that there be a complete data transfer after a granted read request. It only requires that *if* there is a complete data transfer, then an assertion of `read_complete` is necessary. Thus, Assertion 13.10a holds on Trace 13.10(i). If you mean to say that after a granted read request, a complete data transfer followed by an assertion of `read_complete` is necessary, move the complete data transfer ({`data_start ; data[*] ; data_end`}) to the other side of the suffix implication operator, as in Assertion 13.10b. Assertion 13.10b does not hold on Trace 13.10(i).

TIP: Be mindful of where you place the suffix implication operator.

13.1.9 Applying `eventually!` to an implication

Consider the specification: every request (assertion of signal `req`) that is acknowledged (assertion of signal `ack` the following cycle) should receive a grant (assertion of `gnt`) the cycle following the `ack`. We can express this using Assertion 13.11a. Assertion 13.11a holds on Trace 13.11(i) because every acknowledged request is granted. Now let's see what happens if the specification is changed so that an acknowledge can come at any time, rather than only on the next cycle. A beginner might change the first `next` operator to `eventually!`, as in Assertion 13.11b. However, that would be a mistake. To understand why, consider Trace 13.11(ii). Intuitively, our specification should not hold on Trace 13.11(ii) because we have an acknowledged request that does not receive a grant. However, Assertion 13.11b holds on Trace 13.11(ii), contrary to our intuition. For every cycle of Trace 13.11(ii) where `ack` is not asserted, the sub-property `ack -> next gnt` holds (because its else-part defaults to true). In particular, the sub-property `ack -> next gnt` holds at cycle 3. Thus, the

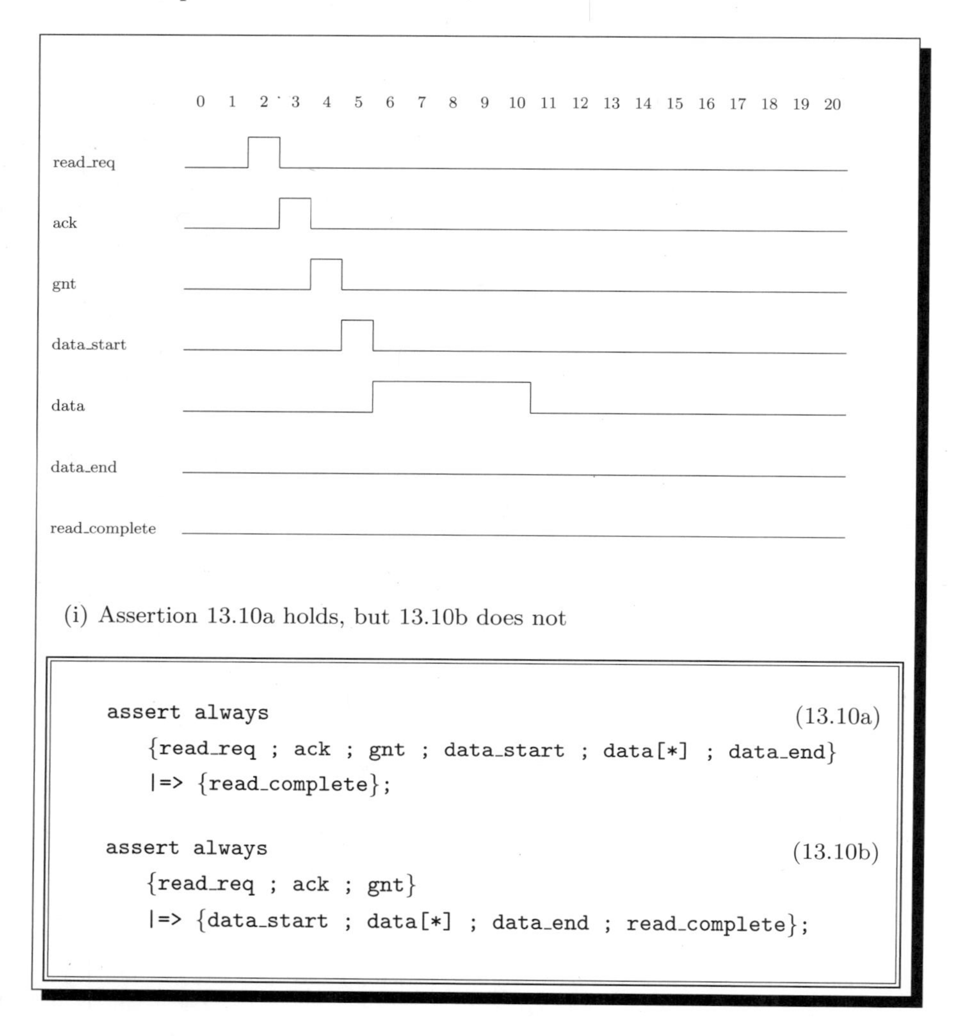

Fig. 13.10: Placement of the suffix implication operator

sub-property `eventually! (ack -> next gnt)` holds at cycle 2 (because the eventuality is fulfilled at cycle 3), and so the assertion holds on the whole trace. The problem is that for any properties `f` and `g`, the property `eventually! (f -> g)` holds if there is at least one future cycle where `f` does not hold. For the same reason, applying `eventually!` to a suffix implication will almost always result in a property that holds on any trace.

One way to fix Assertion 13.11b would be with the `next_event!` operator, as in Assertion 13.11c. Assertion 13.11c states that whenever we see a request, then the next `ack` (which must occur) should be followed by a grant, and does not hold on Trace 13.11(ii).

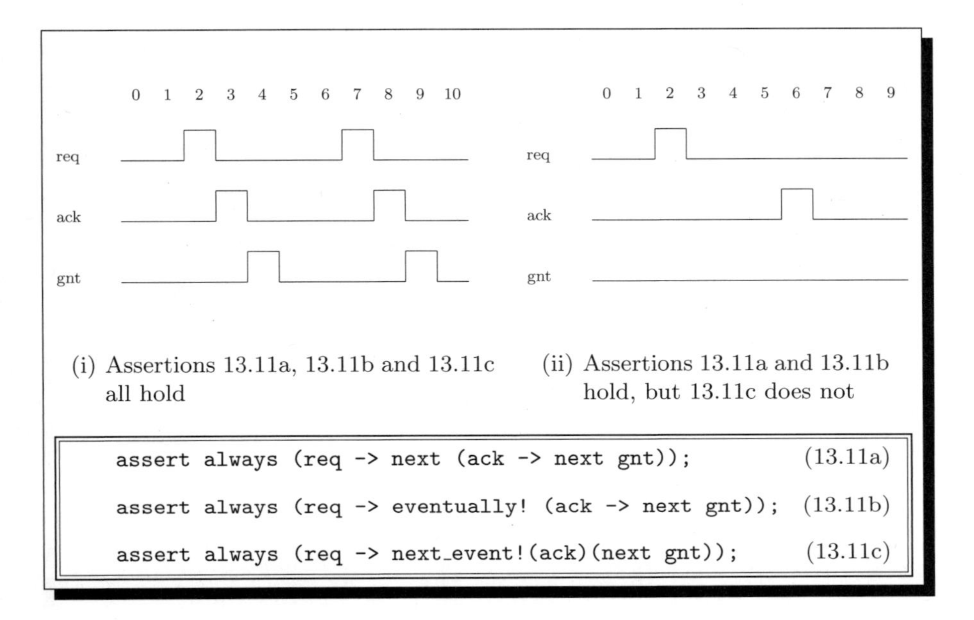

Fig. 13.11: Applying `eventually!` to a logical implication

TIP: Don't apply the `eventually!` operator to a logical implication or to a suffix implication.

13.2 Common errors with `abort`

Beginning users often make mistakes with `abort`. In this section, we examine the most common ones.

13.2.1 Confusing `abort` and `until`

Both the `until` and `abort` operators provide a way to state that some sub-property should hold up to the point where some other sub-property holds. However, they differ in two ways. The first is the way in which the left-hand operand affects the property. Suppose that M is the current cycle and that sub-property `g` holds at some future cycle N. A necessary and sufficient condition for Property 13.12a to hold is that `f` hold on every cycle from M up until N-1. But for Property 13.12b to hold, it is sufficient that `f` hold only at cycle M. This is illustrated in Trace 13.12(i). We can modify Property 13.12b so that it is more similar to Property 13.12a by adding an `always` operator, as shown in Property 13.12c. However, Property 13.12c is not equivalent to Property 13.12a. The two properties mean different things because of the second way in which the `until` and `abort` operators differ.

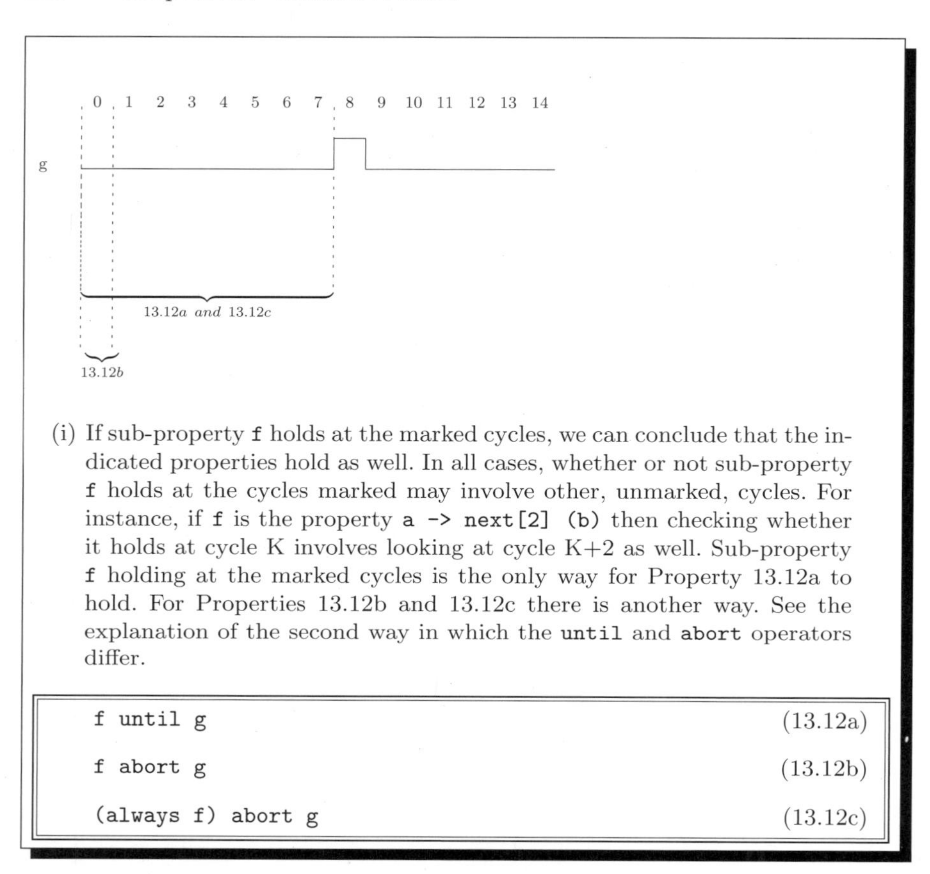

(i) If sub-property `f` holds at the marked cycles, we can conclude that the indicated properties hold as well. In all cases, whether or not sub-property `f` holds at the cycles marked may involve other, unmarked, cycles. For instance, if `f` is the property `a -> next[2] (b)` then checking whether it holds at cycle K involves looking at cycle K+2 as well. Sub-property `f` holding at the marked cycles is the only way for Property 13.12a to hold. For Properties 13.12b and 13.12c there is another way. See the explanation of the second way in which the `until` and `abort` operators differ.

```
f until g                    (13.12a)

f abort g                    (13.12b)

(always f) abort g           (13.12c)
```

Fig. 13.12: Confusing `abort` and `until`

The second way in which the `until` and `abort` operators differ is the way in which the right-hand operand affects the property. The right-hand operand of an `until` property marks the cycle where the left-hand operand is no longer required to hold. However, the cycles affecting whether or not the left-hand operand holds may extend beyond that cycle. For instance, consider Assertion 13.13a. Assertion 13.13a does not hold on Trace 13.13(i) because Assertion 13.13a needs the sub-property `req -> next[2] (ack)` to hold at every cycle between cycle 3, where `fast_mode_start` is asserted, and cycle 12, where `done` is asserted. However, sub-property `req -> next[2] (ack)` does not hold at cycle 11, because `ack` is not asserted at cycle 13. The right-hand side of an `until` operator frees the left-hand side from an obligation to hold in future cycles, but it does not free the left-hand side from the obligation to complete for those cycles prior to the one where the right-hand side holds, even if the cycles needed to complete extend beyond the cycle where the

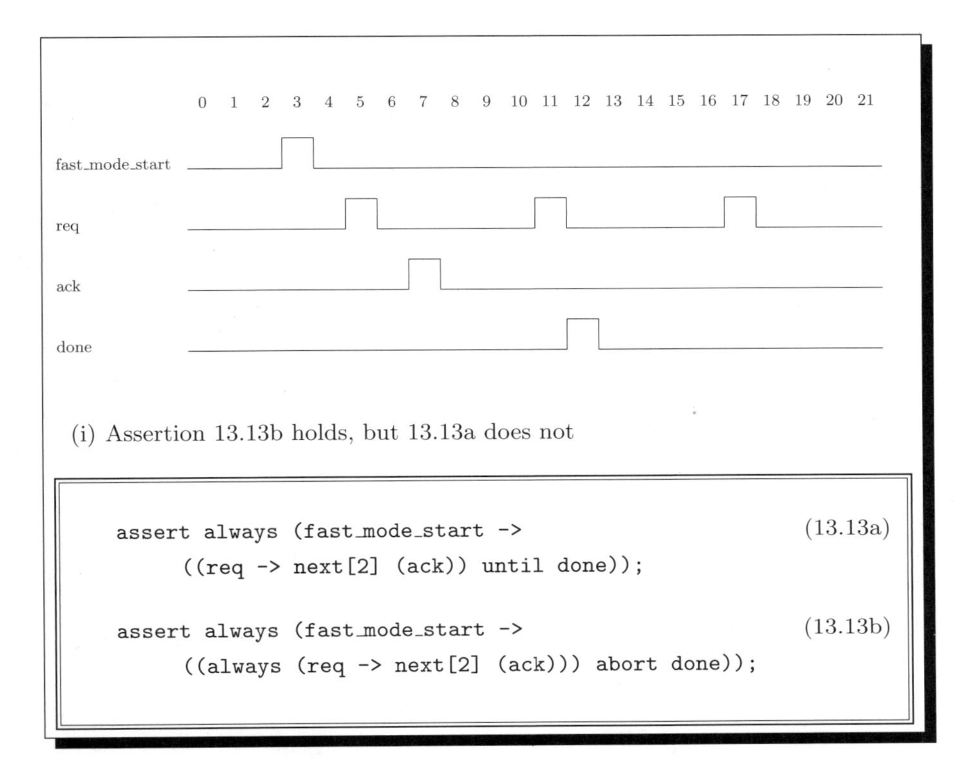

Fig. 13.13: Another way in which `abort` and `until` differ

terminating condition occurs. Thus, the fact that the assertion of `req` at cycle 17 does not get an `ack` is okay, but the fact that the assertion of `req` at cycle 11 does not get an `ack` is not, because it comes before the assertion of `done` at cycle 12.

The right-hand side of an `abort` operator, on the other hand, does free the left-hand side from its obligation to complete. Thus, Assertion 13.13b does hold on Trace 13.13(i). In particular, the assertion of `req` at cycle 11 does not require an assertion of `ack` at cycle 13 because the assertion of `done` at cycle 12 frees the sub-property `always (req -> next[2] (ack))` from the obligation to complete.

TIP: Don't confuse `abort` with `until`.

13.2.2 Confusing `abort` with an "or"

Beginning users often confuse the `abort` operator with the "or" operator. For instance, a beginning user might confuse Property 13.14a and Property 13.14b. But Properties 13.14a and 13.14b are not equivalent: Property 13.14a does

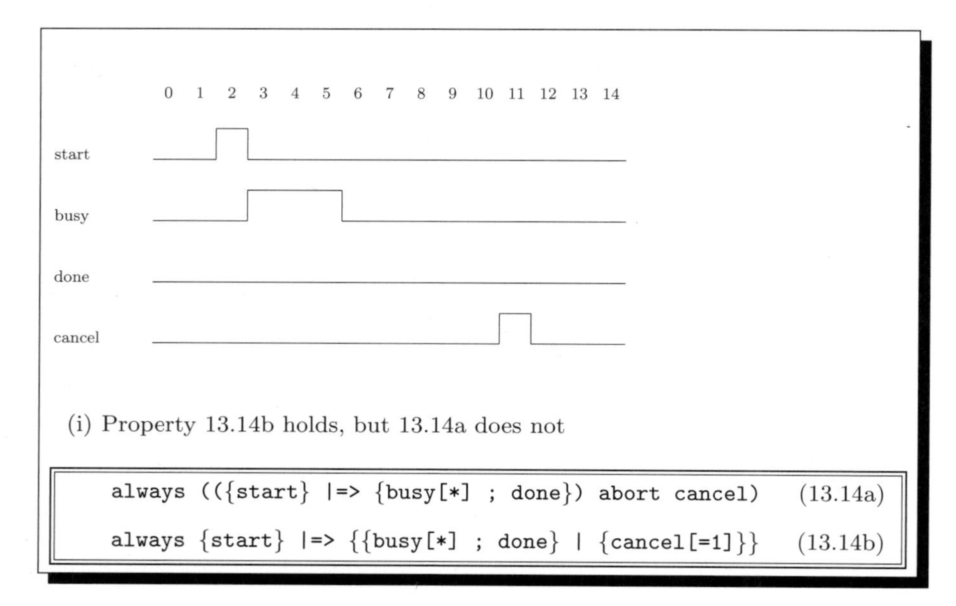

Fig. 13.14: Confusing `abort` with a SERE "or"

not hold on Trace 13.14(i) because something goes wrong with the sub-property `{busy[*] ; done}` before the assertion of `cancel` at cycle 11. Property 13.14b, on the other hand, does hold on Trace 13.14(i) because it says that *either* the sub-property `{busy[*] ; done}` holds, *or* signal `cancel` is asserted once.

TIP: Don't confuse `abort` with an "or" operator.

13.2.3 Incorrectly aborting a `never` property

Consider Assertion 13.15a, which expresses the fact that we must not see three consecutive cycles such that `a` holds on the first, `b` on the second, and `c` on the third. Now let's see what happens when we try to abort this property using signal `reset`. What we would like to express is the fact that the signal `reset` resets the design, that is, that an assertion of signal `reset` should cause all pending requests to hold, and that afterwards the design should again behave according to the aborted sub-property.

A beginner might try Assertion 13.15b. However, this doesn't work. To understand why, consider Trace 13.15(i). Sub-property `{a;b;c} abort reset` holds at cycle 3 of Trace 13.15(i) because sub-property `{a;b;c}` holds at this cycle. Since sub-property `{a;b;c} abort reset` holds at cycle 3, the entire property `never ({a;b;c} abort reset)` does not hold on the trace.

Moving the parentheses in Assertion 13.15b as shown in Assertion 13.15c is not the answer. To understand why, consider Trace 13.15(ii). Assertion 13.15c

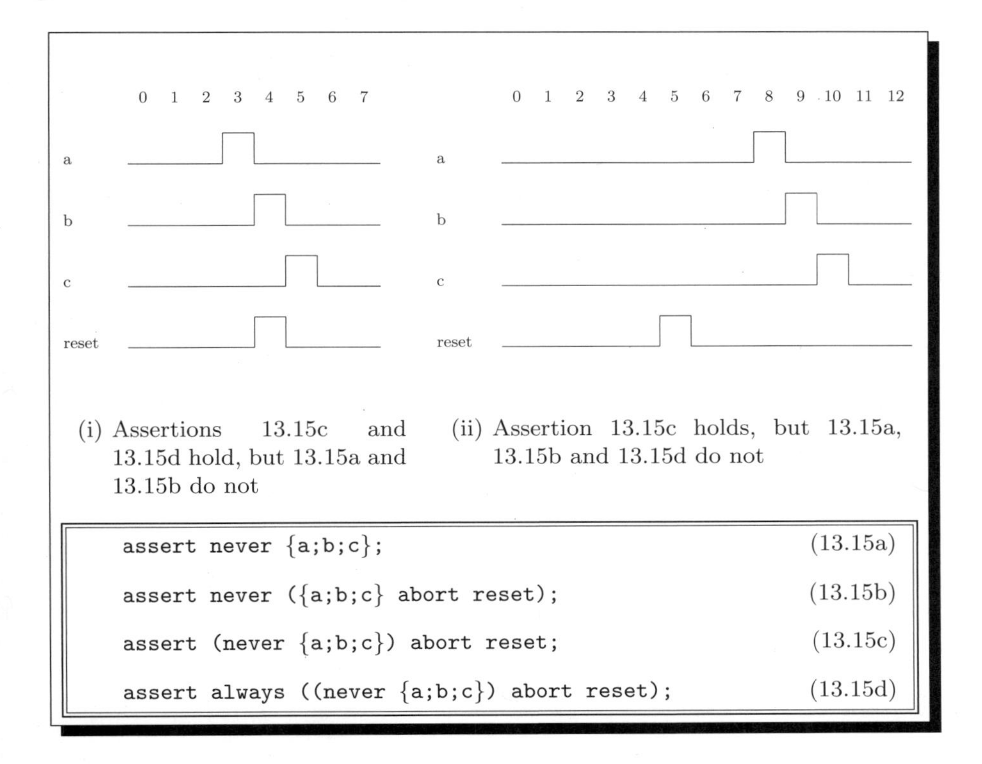


```
assert never {a;b;c};                                  (13.15a)

assert never ({a;b;c} abort reset);                    (13.15b)

assert (never {a;b;c}) abort reset;                    (13.15c)

assert always ((never {a;b;c}) abort reset);           (13.15d)
```

Fig. 13.15: Incorrectly aborting a `never` property

holds on Trace 13.15(ii) because the `abort` operator has been applied to the entire sub-property `(never {a;b;c})`, thus the assertion of `reset` at cycle 5 "saves" the entire trace – after cycle 5, nothing at all is required.

How then should Assertion 13.15a be aborted? As discussed in Chapter 7, the easiest way is to notice that for any property p, `never p` is equivalent to `always never p`. Thus, we can abort Assertion 13.15a by first converting it to that form, then aborting, as in Assertion 13.15d. Assertion 13.15d does not hold on Trace 13.15(ii) because the added `always` operator ensures that the assertion of `reset` at cycle 5 does not have any effect at cycle 6 – at time 6, the whole sub-property `(never {a;b;c}) abort reset` must hold again.

TIP: Be careful when aborting a `never` property.

13.2.4 Incorrectly aborting an initial condition

Some properties refer to initial conditions. For instance, to express the property that the very first assertion of either `req` or `gnt` is an assertion of `req`, we might code Property 13.16a. Property 13.16a contains neither a `never` nor an `always` operator. A beginner might attempt to abort it as shown in

```
req before gnt                                          (13.16a)

(req before gnt) abort rst                              (13.16b)

always (rst -> next ((req before gnt) abort rst))       (13.16c)
```

Fig. 13.16: Aborting a property about an initial condition

Property 13.16b. However, the beginner would be wrong. To understand why, consider that the point of Property 13.16a is to check an initial condition. But the same initial condition that we expect to hold at the start of a trace should hold after reset as well. However, Property 13.16b does not take this into account. As coded, it allows the very first assertion of either `req` or `gnt` after a reset to be a `gnt`. In order to restart after every assertion of `rst`, we add an `always` operator, and an implication the left-hand side of which is `rst` and the right-hand side of which uses `next`, as shown in Property 13.16c. Property 13.16c restarts after every assertion of `rst`. In addition, it aborts the sub-property `(req before gnt)` with `rst` in order to catch the situation in which a second reset comes before the grant is issued. It needs the `next` operator to ensure that the property is not trivially aborted by the same assertion of `rst` that triggered the implication.

TIP: When aborting a property describing an initial condition, remember that initial conditions should hold after reset as well.

13.3 Thinking you are missing a "first match" operator

Suppose we would like to express the following requirement: "on the cycle after the first acknowledge (assertion of `ack`) following a request (assertion of `req`), a data transfer should begin", where a data transfer is indicated by an assertion of `data_start` followed by any number of data beats (assertion of `data`) followed by `data_end`. In other words, the design should behave as shown in Trace 13.17(i). A beginner might start with something like Assertion 13.17a and then get stuck. The requirement is that a data transfer begin after the *first* assertion of `ack` following an assertion of `req`, but Assertion 13.17a requires that one begin after *every* assertion of `ack` following an assertion of `req`, i.e., it requires a data transfer as well for the assertion of `ack` at cycle 10 of Trace 13.17(i). Thus Assertion 13.17a does not hold on Trace 13.17(i), even though Trace 13.17(i) describes the desired behavior. Isn't PSL missing an operator – one that would tell it to stop after seeing the first match of {`req ; [*] ; ack`}? The answer is no. In PSL, you can specify the first assertion of `ack` after `req` by changing the SERE, as shown in Assertion 13.17b. Assertion 13.17b uses the goto repetition operator `[->]`

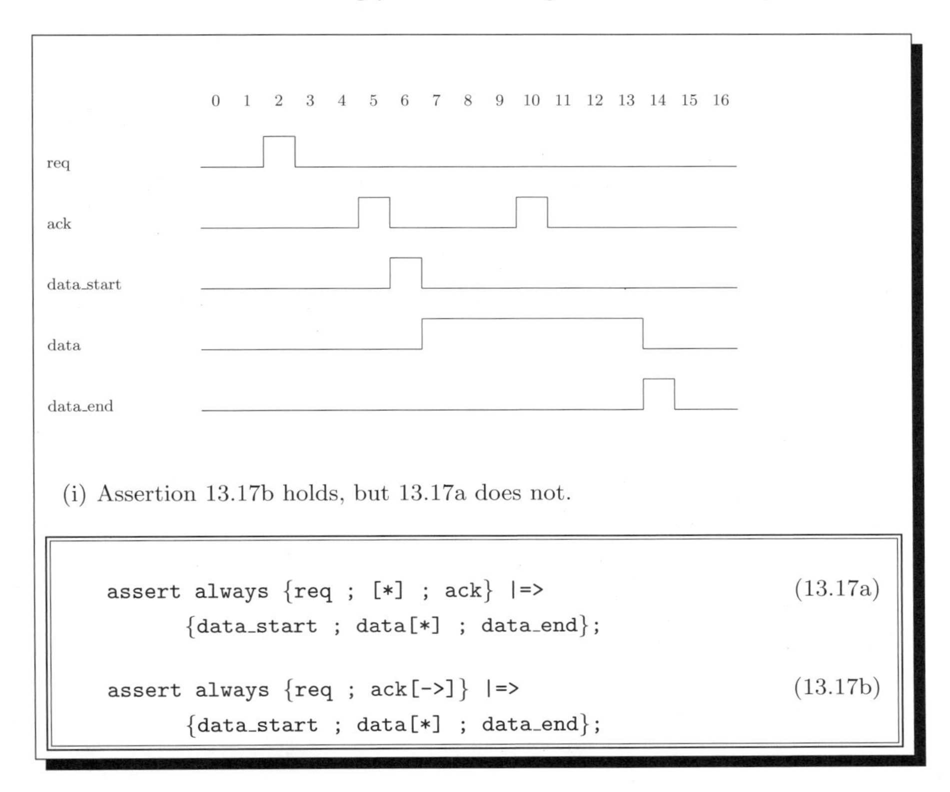

Fig. 13.17: Thinking you are missing a "first match" operator

to indicate that only the first acknowledge after a request should be matched by the SERE. Assertion 13.17b holds on Trace 13.17(i), as desired.

Another mistake beginners often make is thinking that there is something wrong with the right-hand sides of Assertions 13.17a and 13.17b. Couldn't they match longer traces as well, in which `data` continues to be asserted at cycle 14 and forwards, until `data_end` is asserted a second time? And don't we want to prevent this? The answers are 1) that they could indeed match longer traces, and 2) we do not need to prevent this. In order for a suffix implication `{sere1} |=> {sere2}` to hold, every match of `sere1` must be followed by a match of `sere2`. More matches of `sere2` are allowed, but are not required. And since they are not required, there is no need to state explicitly that more are not expected. In summary, the right-hand sides of Assertions 13.17a and 13.17b are fine as they are written.

TIP: If you think you are missing a "first match" operator, try goto repetition (`[->]`) instead.

13.4 Assertions that may say less than you think

Up until now we have concentrated on what an assertion *does say*. In this section, we will take a closer look at what an assertion *does not say*.

13.4.1 "Extraneous" assertions of signals

If the value of a signal is not specified for some cycle, it can have any value at that cycle. This may seem obvious, but it is easy to get confused. For instance, Assertion 13.18a says that whenever an acknowledge is received (assertion of `ack`), it must be followed by an assertion of `start`, some number of busy cycles (assertion of `busy`), and then by an assertion of `done`. It is easy to be lulled into the illusion that Assertion 13.18a describes only simple traces like Trace 13.18(i). However, Assertion 13.18a holds as well on Traces 13.18(ii), 13.18(iii) and 13.18(iv), in which various combinations of signals `ack`, `start`, `busy` and `done` are asserted at "unexpected" times. If we mean that only the behavior of Trace 13.18(i) is allowed, we must explicitly state so, for instance by adding Assertions 13.18b and 13.18c.

As another example, let's start with one of the simple assertions of Section 2.2, repeated in Assertion 13.19a. As we have seen, Assertion 13.19a states that whenever `a` holds, then `b` should hold in the next cycle. Thus, every assertion of `a` should have an associated assertion of `b`, the timing of which is determined by the `next` operator. What Assertion 13.19a *does not* say is the converse: that every assertion of `b` should have an associated assertion of `a`. In other words, "extraneous" assertions of signal `b` not associated with an assertion of signal `a` can be seen, and the assertion will continue to hold. For example, Assertion 13.19a holds on Trace 13.19(i), even though the assertions of `b` at cycles 3 and 11 do not have associated assertions of `a`. If you do not want to see "extraneous" assertions of signal `b`, you must state so explicitly. For example, Assertion 13.19b does not hold on Trace 13.19(i).

TIP: Take care that your assertions say everything that you mean them to.

13.4.2 Specifying a one-to-one correspondence between signals

Let's now examine the simple Assertion 13.20a. Assertion 13.20a states that whenever signal `req` is asserted, signal `ack` should be asserted at some time in the future. What Assertion 13.20a *does not* say is that every request has its own acknowledge. For instance, Assertion 13.20a holds on Trace 13.20(i) because for each of cycles 0, 3, and 6, in which signal `req` is asserted, there is a future assertion of signal `ack`. If this is what we mean (for instance, if the requester is allowed to reissue a request before it is acknowledged), then all is well. But if we intended to specify a one-to-one correspondence between requests and acknowledges, then it is not. If we happen to know that a second request will not be made until an acknowledge is received, then the situation

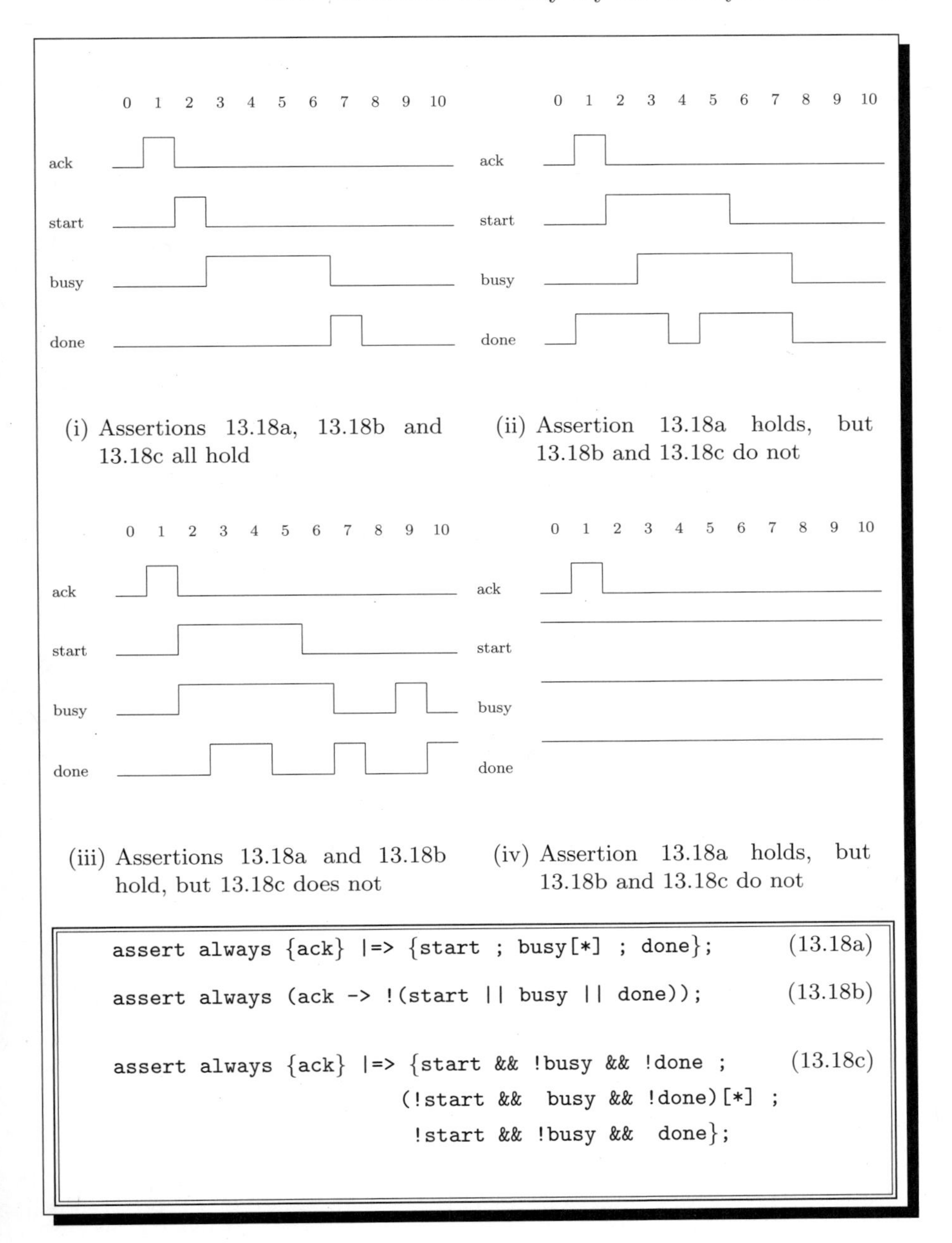

Fig. 13.18: Signals whose value is not specified for every cycle

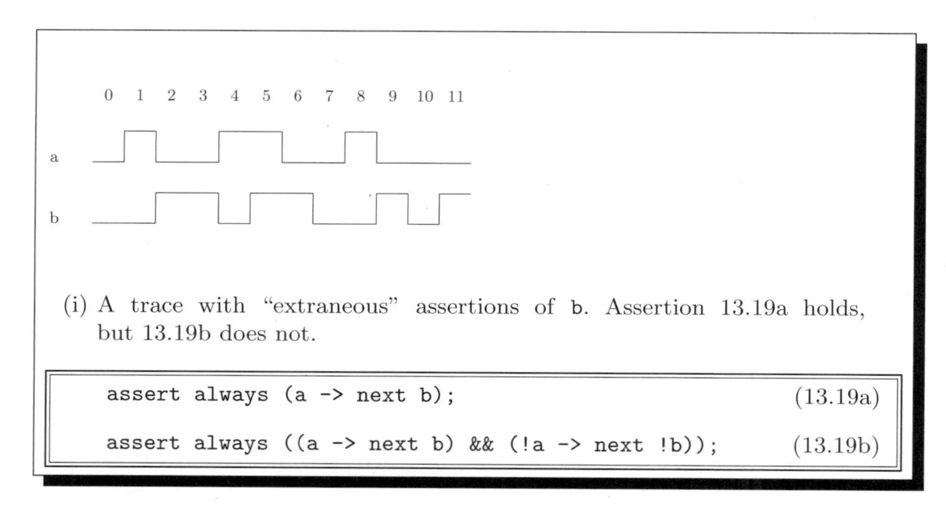

(i) A trace with "extraneous" assertions of `b`. Assertion 13.19a holds, but 13.19b does not.

```
assert always (a -> next b);                              (13.19a)

assert always ((a -> next b) && (!a -> next !b));         (13.19b)
```

Fig. 13.19: "Extraneous" assertions of signals

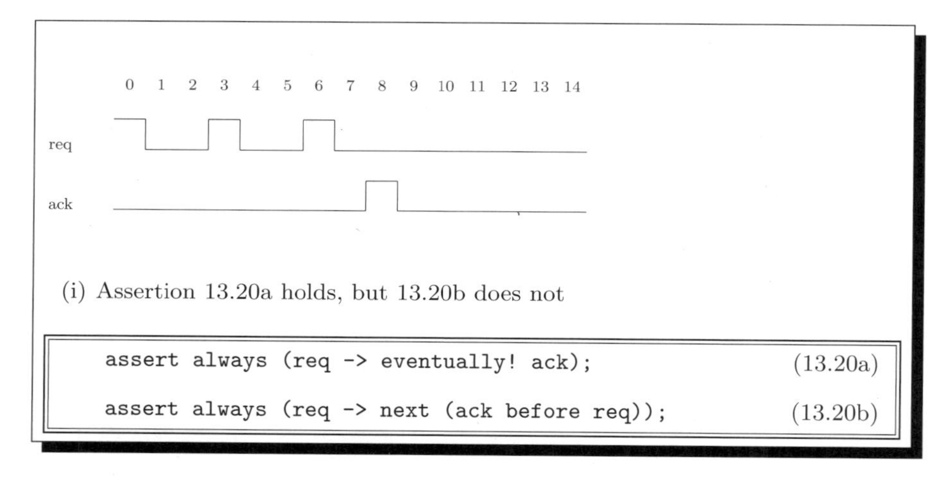

(i) Assertion 13.20a holds, but 13.20b does not

```
assert always (req -> eventually! ack);                   (13.20a)

assert always (req -> next (ack before req));             (13.20b)
```

Fig. 13.20: Specifying a one-to-one correspondence between signals

shown in Trace 13.20(i) should be impossible, and we can write a second assertion that verifies this – for instance Assertion 13.20b.

However, if it is possible for multiple assertions of `req` to come before a single assertion of `ack`, and if in addition each such request should receive its own acknowledge, then Assertion 13.20a is not suitable for our purposes. What we need is an assertion that ensures that every request receives its own acknowledge. A way to do that in PSL is to use the modeling layer to tag the requests and acknowledges, then to write an assertion that "matches" each request with its acknowledge. For instance, the vunit `one_to_one_example1`

```
vunit one_to_one_example1 {
   reg [3:0] count, req_count, ack_count;
   initial begin
      count[3:0] <= 4'b0;
      req_count[3:0] <= 4'b0;
      ack_count[3:0] <= 4'b0;
   end;

   always @(posedge clk)
      if (req & !ack)
        count[3:0] <= count[3:0] + 1;
      else if (!req & ack)
        count[3:0] <= count[3:0] - 1;

   always @(posedge clk)
      if (req)
        req_count[3:0] <= req_count[3:0] + 1;

   always @(posedge clk)
      if (ack)
        ack_count[3:0] <= ack_count[3:0] + 1;

   assert always ((count[3:0]==15) -> !(req & !ack));
   assert always ((count[3:0]==0) -> !ack);

   assert forall i in {0:15}:
      always ((req && req_count[3:0]==i) ->
          eventually! (ack && ack_count[3:0]==i));
}
```

Fig. 13.21: Specifying a one-to-one correspondence between reqs and acks

of Figure 13.21 tags each request with the value in `req_count`, and each acknowledge with the value in `ack_count`. It does so as part of the modeling code, so these tags do not become part of the design – they are used only when checking the properties. Note that by choosing the width of the signals `req_count` and `ack_count`, we made an assumption on the maximum number of requests that can be outstanding at any one time. This is not a problem, because obviously our hardware must also make such an assumption.

In order to be extra careful, we check our assumption by counting the number of outstanding requests in auxiliary signal `count[3:0]`. Signal `count[3:0]` is incremented if a request is issued without a simultaneous acknowledge, and is decremented if an acknowledge is received without an additional request being issued (we have assumed that a request and an acknowledge can come simultaneously).

The first assertion of vunit one_to_one_example1 checks that we have selected a large enough bound for the number of outstanding requests. If there are 15 outstanding requests and we try to record another one, our counter will roll over. This assertion detects that case, which indicates either that our modeling code needs to be modified in order to code for a larger bound, or that there is a problem in the design or in the environment which is causing a greater number of outstanding simultaneous requests than was expected.

The second assertion checks that there are no extraneous acknowledges – i.e., that if count is zero, no acknowledge is issued. This of course assumes that an acknowledge cannot be issued immediately – that is, a request must wait at least a single cycle for its acknowledge. If it was possible for an acknowledge to be issued in the same cycle as its request, this assertion would not be correct. Rather, we would have to code an assertion saying that if count is zero, and no request is being made this cycle, then no acknowledge is issued.

Finally, the third assertion checks that if a request is issued, eventually an associated acknowledge will be issued.

The `eventually!` operator is not the only operator into which a one-to-one correspondence is sometimes read when none is there. As a second

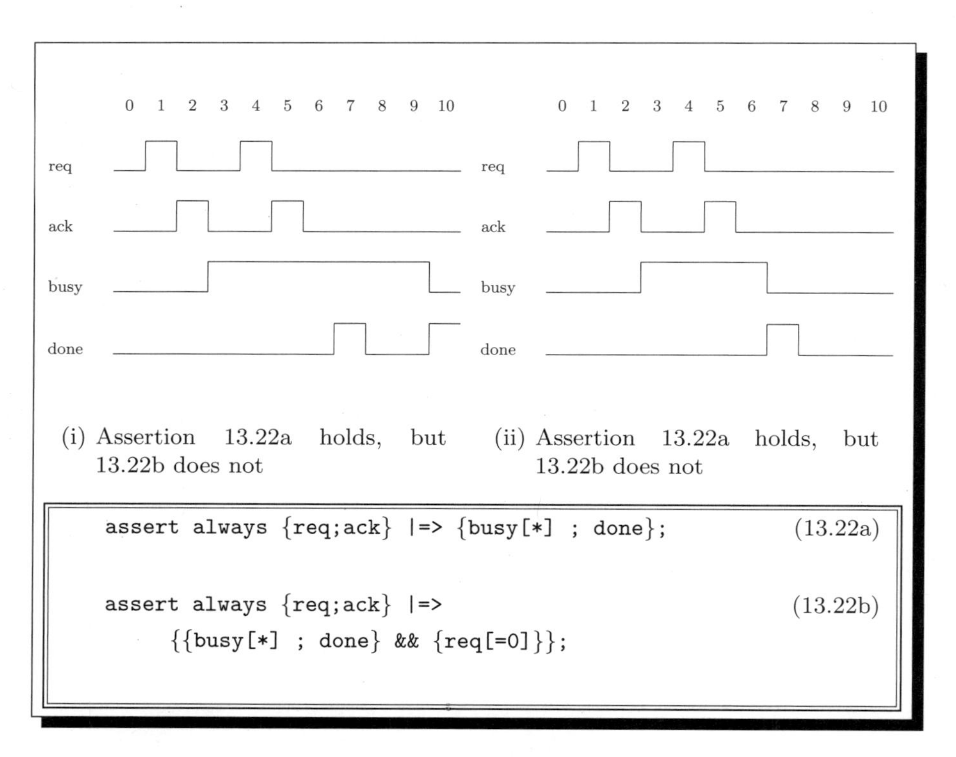

Fig. 13.22: Specifying a one-to-one correspondence between sequences of cycles

```
vunit one_to_one_example2 {
   reg [3:0] count, req_count, done_count;
   initial begin
      count[3:0] <= 4'b0;
      req_count[3:0] <= 4'b0;
      done_count[3:0] <= 4'b0;
   end;

   always @(posedge clk)
      if (ended({req;ack}))
        count[3:0] <= count[3:0] + 1;
      else if (ended({busy[*];done}))
        count[3:0] <= count[3:0] - 1;

   always @(posedge clk)
      if (ended({req;ack}))
        req_count[3:0] <= req_count[3:0] + 1;

   always @(posedge clk)
      if (ended({busy[*];done}))
         done_count[3:0] <= done_count[3:0] + 1;

   assert always ((count[3:0]==15) -> !ended({req;ack}));
   assert always ((count[3:0]==0) -> !ended({busy[*];done}));

   assert forall i in {0:15}:
      always {req ; ack && req_count[3:0]==i} |=>
                 {busy[*] ; done && done_count[3:0]==i};
   assert never (ack & done);
}
```

Fig. 13.23: One-to-one correspondence between req-ack pairs and their completions

example, consider Assertion 13.22a on Trace 13.22(i). Assertion 13.22a holds on Trace 13.22(i), but the second assertion of `done` has nothing to do with it – Assertion 13.22a holds on Trace 13.22(ii) as well. If we mean to allow the behavior of Trace 13.22(ii), all is well. However, if we do not, then as in the previous example, there are two possibilities.

Either our design is such that we do not expect another `req-ack` pair before seeing an assertion of `done`, or we allow another `req-ack` pair before seeing an assertion of `done` for the first pair, but we expect the second pair to see a second assertion of `done`. In the first case, we can use Assertion 13.22b to describe our design. In the second case, we will need the modeling layer. Figure 13.23 shows vunit `one_to_one_example2` that tags the outstanding requests (up to a maximum of 15) and completions and ensures that each request gets its own assertion of `done`. For simplicity, we have assumed that `done` will not be asserted together with `ack`, and we have added an assertion to check this.

The vunit of Figure 13.23 is similar to the vunit of Figure 13.21, with two main differences. The first is that the counters are triggered by calls to `ended()` on the SEREs {`req;ack`} and {`busy[*] ; done`}, and the second is the presence of the additional assertion verifying our simplifying assumption that signal `done` cannot be asserted together with `ack`. If that simplifying assumption does not hold, then our counter needs to be modified to check for this condition, in the same way that the counter in Figure 13.21 takes into account the case that `req` and `ack` can be asserted together.

TIP: Don't be fooled into reading a one-to-one correspondence where there is none.

13.5 Assertions that may say more than you think

In the previous section, we examined the problem of specifying a one-to-one correspondence between signals. Another way that a beginner might try to specify such a thing leads to a property that says more, rather than less, than you might think.

As an example, let's return to the issue of a one-to-one correspondence between requests (assertions of `req`) and acknowledges (assertions of `ack`), in the case that a new request cannot come until the previous one has been acknowledged. A beginner might try Property 13.24a. However, the beginner would be wrong. To see why, examine Trace 13.24(i). Trace 13.24(i) contains exactly one request and one acknowledge. Therefore, we would like it to hold. However, Property 13.24a does not hold on Trace 13.24(i): the `always` operator requires that sub-property `req before ack` hold at every cycle. In particular, it should hold at cycle 4. However, starting from cycle 4 and looking forward (which is how to treat a current cycle), we see that `ack` is asserted before `req`, rather than the other way around. Property 13.24a applies the `always` operator too zealously, with the result that the property says more than was intended.

If we want to ensure that no acknowledge comes without a corresponding request, we can code Property 13.24b. The left-hand side of Property 13.24b (`req before ack`) takes care of the very first acknowledge. It holds only if the very first acknowledge is preceded by a request. The right-hand side of Property 13.24b (`always (ack -> next (req before ack))`) takes care of the rest of the acknowledges. It holds only if all of the rest of the acknowledges are preceded by requests.

Note that Property 13.24b is not in the simple subset. To overcome this, split it into the two Properties 13.24c and 13.24d.

TIP: Don't be fooled into saying more than you mean.

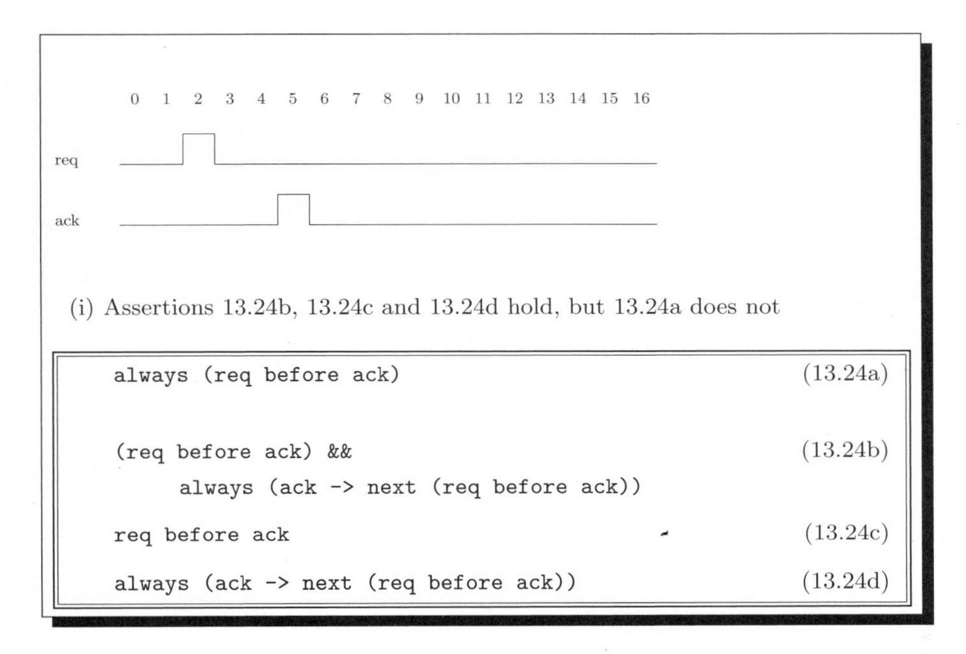

Fig. 13.24: Assertions that may say more than you think

13.6 A False "equivalence"

In Chapter 11 we examined some common equivalences between PSL properties, such as the equivalence between `always p` and `never !p`, for any PSL property `p`. Beginning users often discover false "equivalences", that is, sets of properties that seem to be equivalent but are not. In this section we examine one of those false "equivalences" in detail.

Given a named SERE `s`, it is easy to think that properties `next_a[1:2] (s)` and `{'true ; s[*2]}` might express the same requirement. The reason is first that for a Boolean `b`, the equivalence between `next_a[1:2] (b)` and `{'true ; b[*2]}` does hold (as we saw in Chapter 11). And second, if SERE `s` lasts for exactly one cycle, the equivalence holds for `s` as well. However, if `s` lasts for longer than a single cycle, then the two are different.

The difference stems from the different way in which `next_a[1:2]` and `[*2]` need to see two occurrences of their respective operands. While `next_a[1:2]` needs to see its operand happen at the next cycle and at the one following that, `[*2]` needs to see two occurrences, the second of which starts only after the first has finished.

For instance, for the definition of `s2` in Figure 13.25, Property 13.25a holds if `s2` holds on the next cycle and the one following. For this to happen, the SERE `{a;b}` should hold at cycle 1 and at cycle 2. Thus Property 13.25a holds on Trace 13.25(i), but not on Trace 13.25(ii). However, Property 13.25b

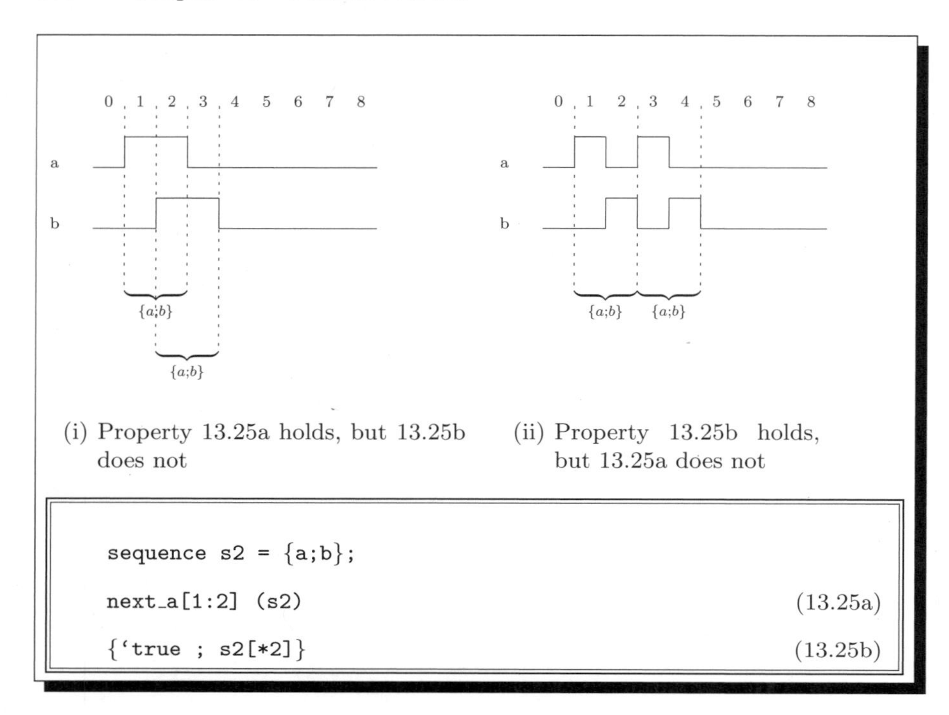

Fig. 13.25: An illustration of the difference between Properties 13.25a and 13.25b

holds if starting at the next cycle, `s2` holds, and then holds again after the first `s2` has completed. For this to happen, the SERE {`a;b`} should hold at cycle 1 and at cycle 3. Thus, Property 13.25b holds on Trace 13.25(ii), but not on Trace 13.25(i).

TIP: When using a SERE in conjunction with non-SERE operators like `next`, make sure you understand what you are doing.

13.7 Unspoken assumptions

Consider the `within` example of Chapter 5, repeated here in Figure 13.26. We were trying to state that although the normal behavior of the design is to complete a high priority request first, even if there is a pending low priority request that started before it, this is prohibited if signal `no_nesting` is asserted when the low priority request is issued. In other words, that the situation in Trace 13.26(i) is not allowed. If there is guaranteed to be no pipelining of low priority requests, Assertion 13.26a says all that we want it to. However, if low priority requests can be pipelined, then Assertion 13.26a is not enough.

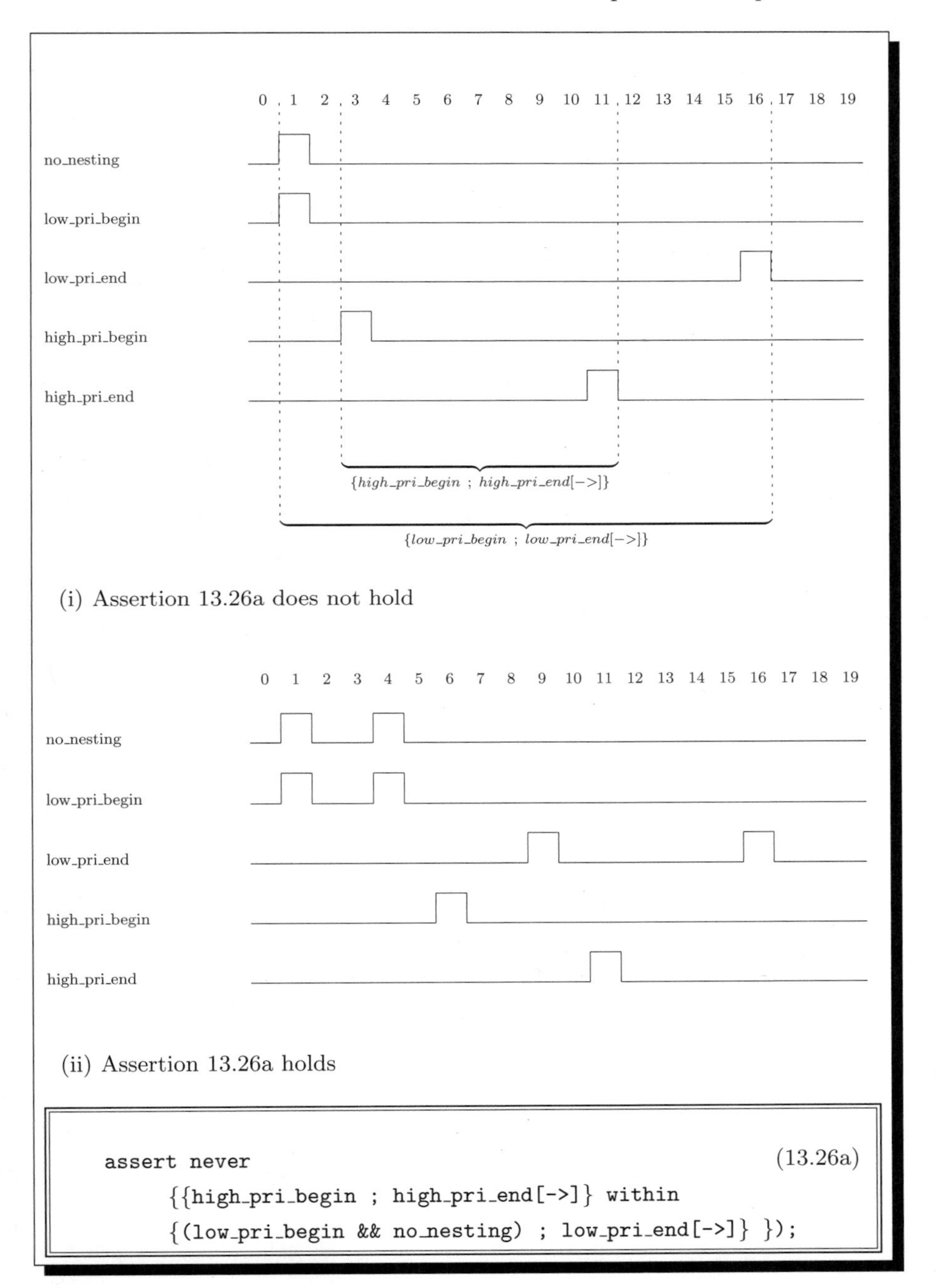


```
assert never                                                  (13.26a)
     {{high_pri_begin ; high_pri_end[->]} within
     {(low_pri_begin && no_nesting) ; low_pri_end[->]} });
```

Fig. 13.26: Unspoken assumptions

To see why, consider Trace 13.26(ii). We were trying to express the fact that because of the assertion of `no_nesting` at cycle 4, the low priority request begun at cycle 4 should complete before the high priority request begun at cycle 6. However, Assertion 13.26a holds on Trace 13.26(ii), contrary to what we were trying to achieve.

The problem stems from the fact that Assertion 13.26a was written for the simple case, without taking the pipelining into consideration. As written, the assertion of `low_pri_end` at cycle 9 is associated with the assertion of `low_pri_begin` at cycle 1, as it should be, but also with the assertion of `low_pri_begin` at cycle 4, which is not what was meant. Thus, it does not catch the violation of the specification we are trying to express, in which a match of {`high_pri_begin ; high_pri_end[->]`} occurs within the sequence of cycles 4 through 16 corresponding to the beginning and end of the second low priority request. The solution is similar to the solutions seen in Section 13.4.2, and is left as an exercise for the reader.

TIP: Try to think of a trace that is bad, but adheres to your property. Such an exercise will typically uncover holes in your specification.

14

Multiply-clocked Designs

In Chapter 6, we examined singly-clocked designs and multiply-clocked designs where all signals in a property are clocked on the same clock. Suppose now that we have the case shown in Figure 14.1. That is, we want to specify that signal `b` is a latched version of signal `a`, where signals `a` and `b` are clocked with different clocks. Obviously, we will have to write a property that contains multiple clocks. As we have seen in Chapter 6 for the case of singly clocked designs, the issue of whether the design is edge-triggered or level-sensitive is one of implementation – the clocking can be performed in both cases by selecting PSL cycles in which the signals are guaranteed to be at a steady state.

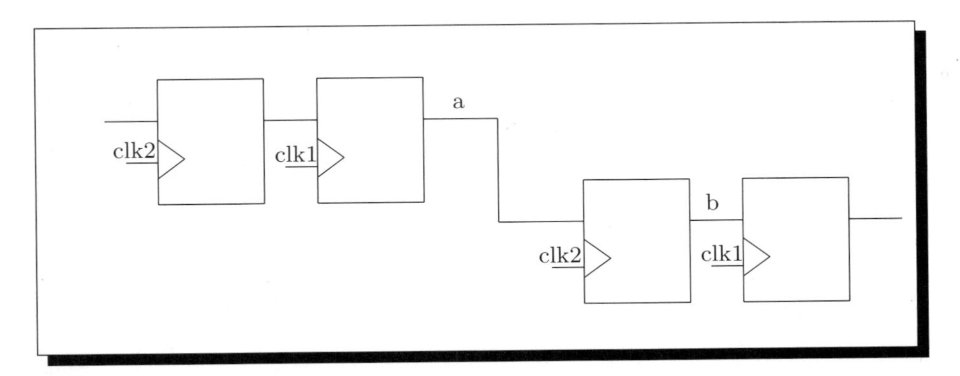

Fig. 14.1: A level-sensitive design, where `a` and `b` are clocked on different clocks

Let's assume that the latches of Figure 14.1 are either edge-triggered on the rising edge of the clock, or level-sensitive to the high value of the clock. The solution in the case of the falling edge or sensitivity to the low value of the clock is of course similar. If the clocks are well-behaved – that is, if they are interleaved – then signals `a` and `b` might behave as shown in Trace 14.2(i).

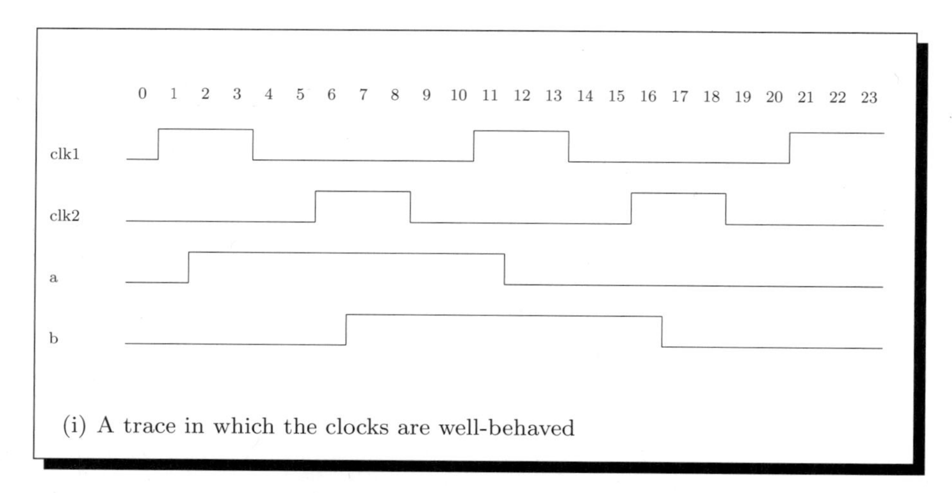

Fig. 14.2: Well-behaved clocks

Intuitively, we'd like to look at pairs of cycles (2,7), (12,17), etc. Which PSL cycles should our clock operator filter in order to see these pairs? We would like to filter the cycles of `@rose(clk1)` for signal `a`, and those of `@rose(clk2)` for signal `b`, as shown in Traces 14.3(i) and 14.3(ii), because in a glitch-free design the value of `a` at a rising edge of `clk1` is guaranteed to be the same as the value latched by the previous tick of `clk1`, and the same for `b` on `clk2`. Thus, instead of cycles 2 and 7, we will see equivalent cycles 11 and 16, and instead of cycles 12 and 17, we will see equivalent cycles 21 and 26 (not shown), etc. Doing so will work in most cases. However, notice that choosing `@rose(clk1)` and `@rose(clk2)` means that we have a pair of cycles, (1,6), which do not have a corresponding pair in our intuitive view. In a glitch-free design, these are the initial values of signals `a` and `b`. Let's assume for the moment that this extra pair of cycles does not bother us (because, for instance, we are not interested in the initial values). We will get back to the issue of how to clock in the case that this is not so later on. Assuming that the initial values of `a` and `b` are not an issue at the moment, and that therefore we want to filter the cycles of `rose(clk1)` for signal `a` and those of `rose(clk2)` for signal `b`. We can do this in PSL as shown in Property 14.3a.

Now, suppose that we are concerned about the initial values of signals `a` and `b`. Assume that the design behaves as shown in Trace 14.4(i). In Trace 14.4(i), signal `b` is still a latched version of signal `a`, but the initial value of `a` is 1 while the initial value of `b` is 0. Thus, Property 14.4a does not hold on Trace 14.4(i). If we intend to catch this as an error, all is well. But if we intend to say only that `b` is a latched version of `a`, but do not wish to say anything about the initial values, we have a problem – there is an "extra" pair of PSL cycles, (1,6), that does not adhere to our property, even though that is not our intention. We can solve this problem in one of two ways: we

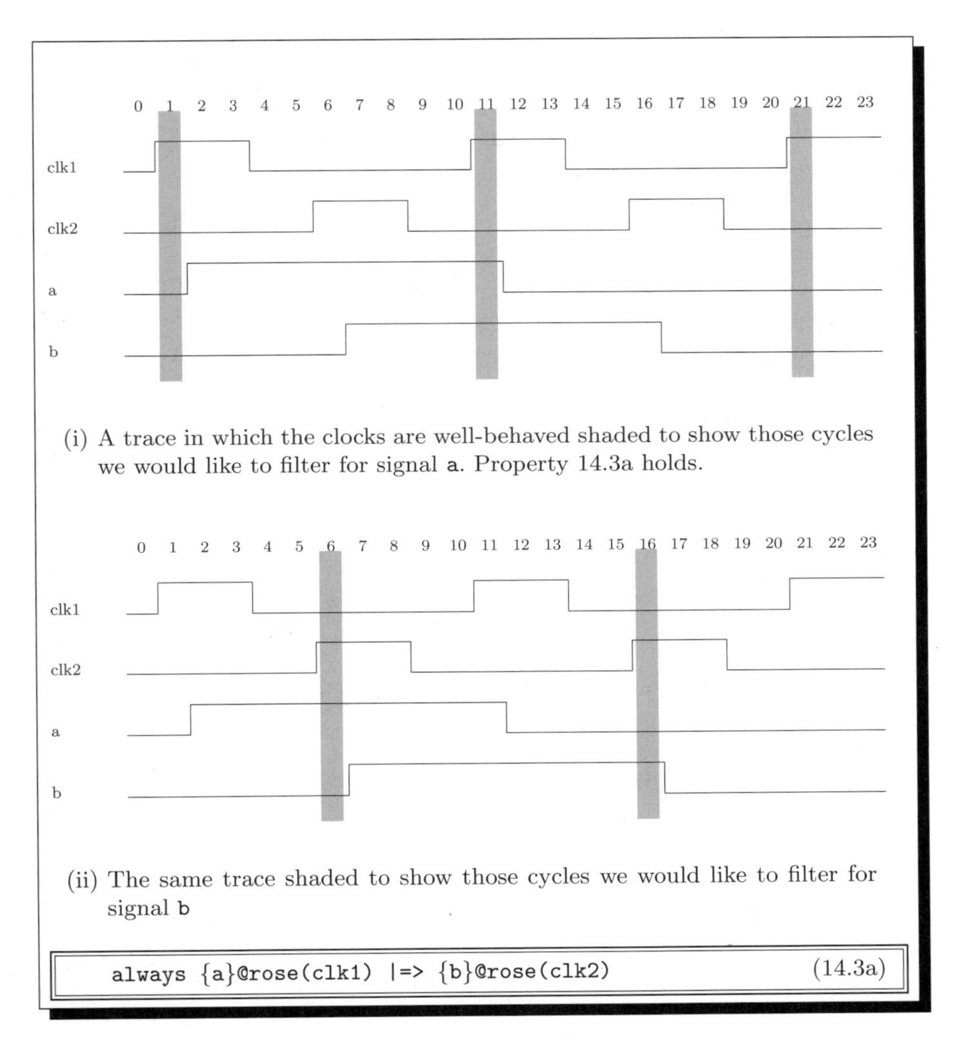

(i) A trace in which the clocks are well-behaved shaded to show those cycles we would like to filter for signal a. Property 14.3a holds.

(ii) The same trace shaded to show those cycles we would like to filter for signal b

```
always {a}@rose(clk1) |=> {b}@rose(clk2)                    (14.3a)
```

Fig. 14.3: Clocking well-behaved clocks

can either choose different cycles on which to clock signals a and b, or we can change the unclocked property in order to skip the first pair of cycles (1,6). If we choose to change the property, we can modify Property 14.4a as shown in Property 14.4b. The [*1] in Property 14.4b has the effect of skipping the first pair of cycles (1,6) seen by the corresponding Property 14.4a.

While it works, the solution of modifying the property using [*1] is complicated, in that a property with a different format than that of Property 14.3a would have to be modified in a different way. A cleaner solution is to choose different cycles on which to clock the property. We previously argued that

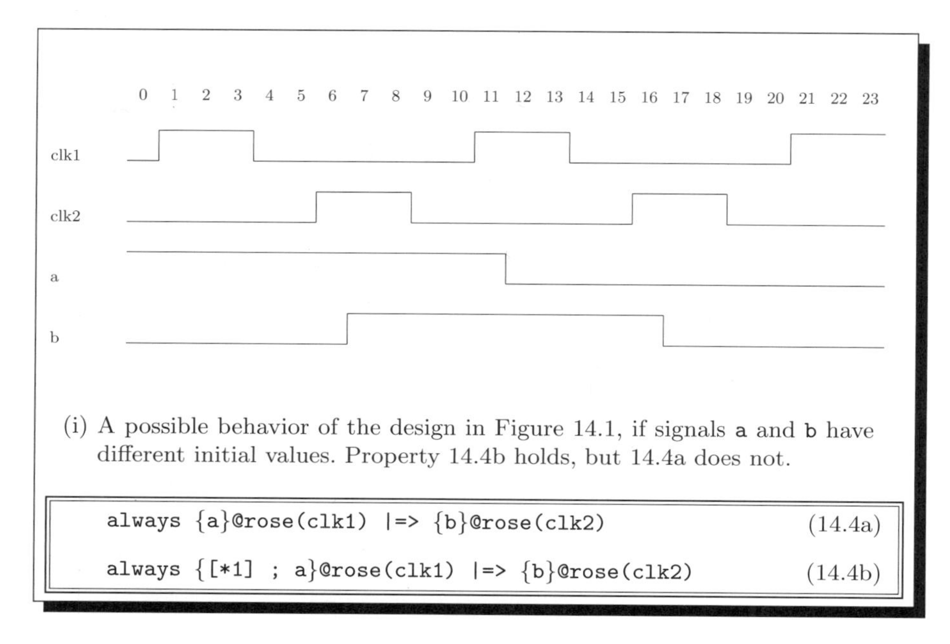

(i) A possible behavior of the design in Figure 14.1, if signals a and b have different initial values. Property 14.4b holds, but 14.4a does not.

```
always {a}@rose(clk1) |=> {b}@rose(clk2)                    (14.4a)

always {[*1] ; a}@rose(clk1) |=> {b}@rose(clk2)             (14.4b)
```

Fig. 14.4: A multiply-clocked design, where the initial values are important

cycle pair (11,16) can stand in for pair (2,7), because signal a is guaranteed not to change between cycles 2 and 11, and signal b is guaranteed not to change between cycles 7 and 16. Thus, we could equivalently have chosen any of the pairs (x,y) such that $2 \leq x \leq 11$ and $7 \leq y \leq 16$. At least one of these pairs, (4,9), is easily filtered in PSL by `@fell(clk1)` and `@fell(clk2)`. And, as shown in Traces 14.5(i) and 14.5(ii), clocking on the falling edge of `clk1` and `clk2` has the advantage of skipping over the initial values of signals a and b that caused us problems in Property 14.4a. Thus, paradoxically, in a multiply-clocked property using latches triggered on the rising edge of clocks `clk1` and `clk2`, or latches sensitive to their high value, we can easily express our specification as shown in Property 14.5a.[1]

We've seen that for multiply-clocked properties, clocking in the opposite direction to that of the latches can provide a solution. Could we do the same in a singly-clocked property? That is, given a singly-clocked design using latches that are triggered on the rising edge or sensitive to the high value of clock `clk`, could we clock using `@fell(clk)`? The answer is no. To understand why,

[1] Note that the problem we have examined and its solution are dependent on the fact that the first clock edge in the trace is a rising edge of `clk1`. If the design had started off with a falling edge of `clk1`, or with a rising or falling edge of `clk2`, the precise problem and solution would be slightly different. Using reasoning similar to that we have described, you should easily be able to decide on a clocking scheme appropriate for your particular design.

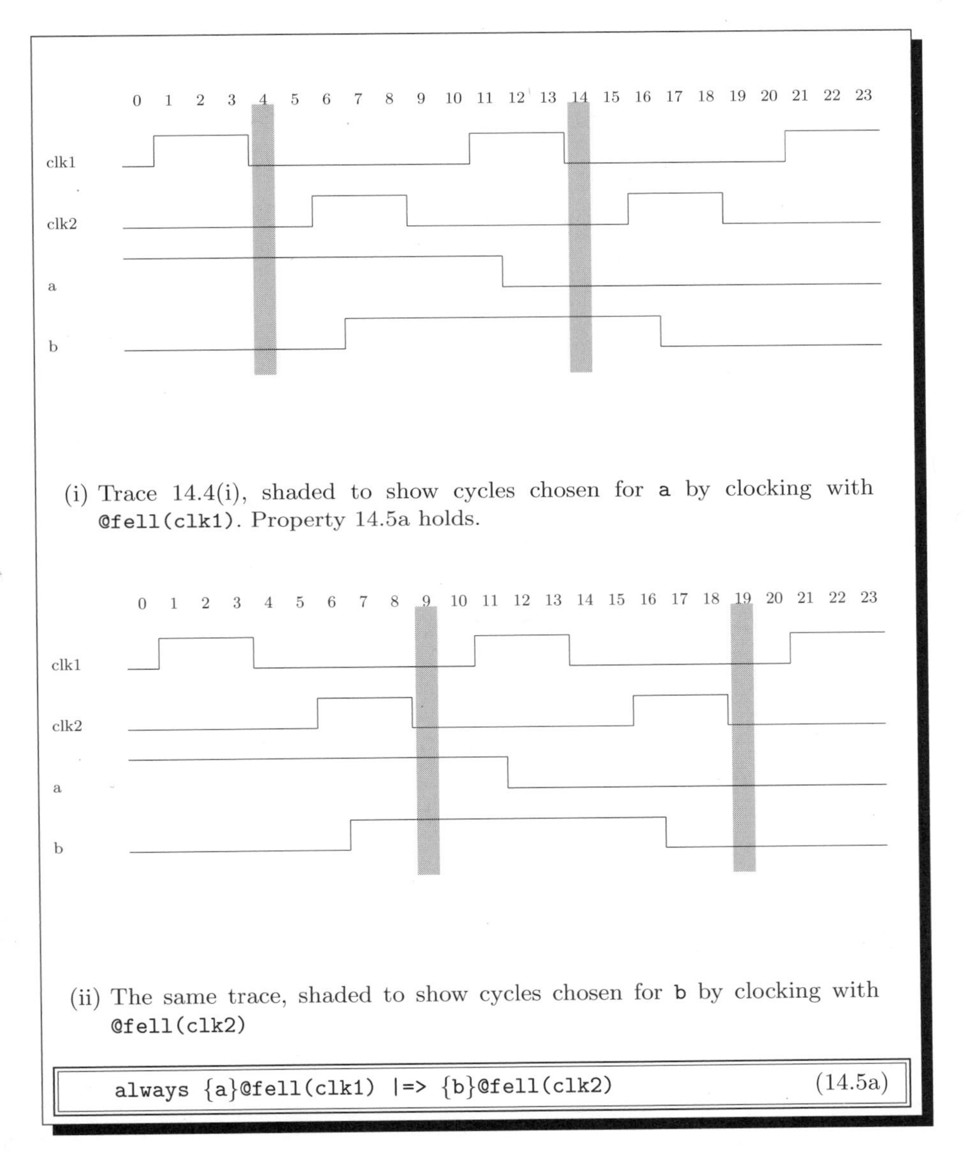

(i) Trace 14.4(i), shaded to show cycles chosen for `a` by clocking with `@fell(clk1)`. Property 14.5a holds.

(ii) The same trace, shaded to show cycles chosen for `b` by clocking with `@fell(clk2)`

```
always {a}@fell(clk1) |=> {b}@fell(clk2)          (14.5a)
```

Fig. 14.5: Clocking a multiply-clocked design when the initial values are important

consider that the reason that we used `@fell(clk)` to describe the behavior of Trace 14.5(i) was that we wanted a way to skip over the first pair of clock cycles, because they represent initial values. In a singly-clocked design, the first clock cycle, not the first pair, represents the initial values, so in a singly-clocked design we do not want to skip over the first pair – we will miss something important. Thus, we need to clock such designs using `@rose(clk)` if they are edge-triggered on the rising edge or level-sensitive to the high value of the clock, and using `@fell(clk)` otherwise.

14.1 Declaring auxiliary clocks

We've seen that in a multiply-clocked property, we can clock in the direction "opposite" to that of the latches. While this works, it can be confusing to a person trying to understand the specification. Another option is to add a signal to the design, which is asserted in those cycles which we would like to use to check our properties. For instance, suppose that we know that by the last (PSL) cycle of the clock, signals will be stable. We can add signal `pulsedclk1` and `pulsedclk2` to the design, with the behavior shown in Trace 14.6(i). Then, we can clock our property with it as shown in Property 14.6a.

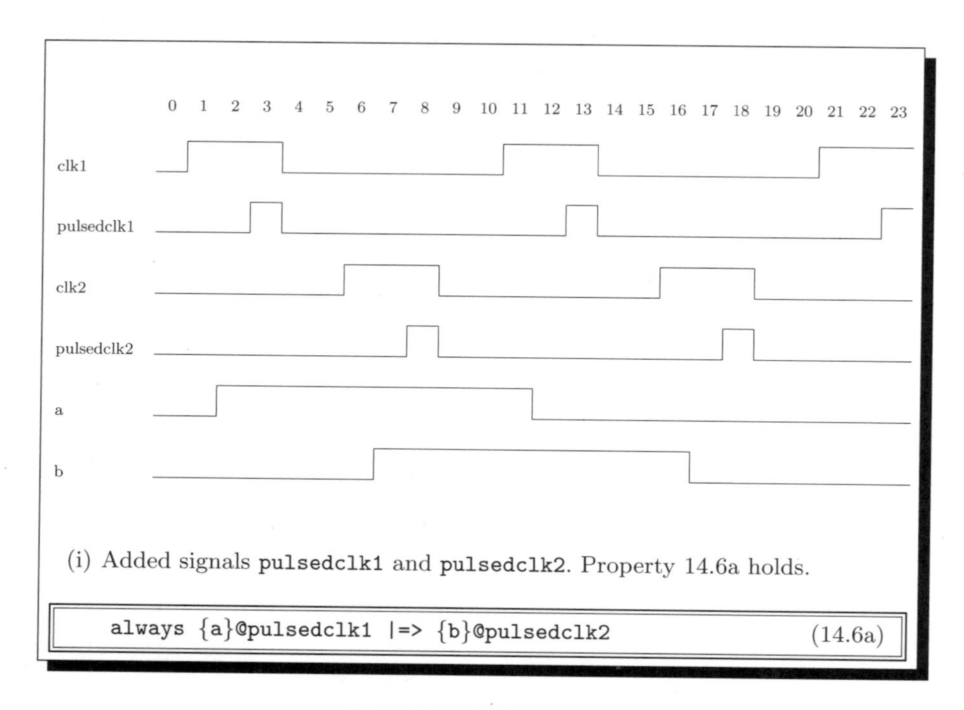

(i) Added signals `pulsedclk1` and `pulsedclk2`. Property 14.6a holds.

```
always {a}@pulsedclk1 |=> {b}@pulsedclk2                    (14.6a)
```

Fig. 14.6: Pulsed clocks

Since this solution avoids using `@fell(clk1)` and `@fell(clk2)` in a design triggered on the positive edge or sensitive to the high value of the clock, it can be used to improve the readability of a PSL specification of such a design.

14.2 Multiply-clocked designs vs. properties

In a singly-clocked design, all properties will be singly-clocked. In a multiply-clocked design, however, there might be some properties that are singly-clocked, while others are multiply-clocked. Obviously, those that are multiply-clocked should be clocked as discussed previously – in the direction of the latches if we want to include the first pair of clock cycles, or in the direction opposite to that of the latches or using auxiliary signals like `pulsedclk1` and `pulsedclk2` of Figure 14.6 if we do not. What about singly-clocked properties in a multiply-clocked design? Should they be clocked in the direction of the latches, as if they came from a singly-clocked design, or possibly in the direction opposite to the latches like some multiply-clocked properties of the design? The answer can be found by thinking about what we want to achieve. In a design with two clocks, in which we want to express the specification "two consecutive requests (assertion of signal `req`) are not allowed" and clock it on the rising edge of signal `clk1`, do we want to skip over the first pair of clocked edges or not? The answer, of course, is no, because the initial value of `req` is latched on the first rising edge of its clock, thus in such a case we should clock with `@rose(clk1)`. Decide how to clock your property according to what you want to say, not according to how many clocks there are in the design.

14.3 Nesting of clocks

Clocks can be nested. For instance, Property 14.7a is a legal property. When clocks are nested, they do not accumulate. Rather, an inner clock takes precedence over an outer clock. Using the rewrite rules of Appendix B, it is easy to see this: rewriting `b@clk1@clk2` gives `b@clk1`. From the rewrite rules, we get that Property 14.7a is equivalent to Property 14.7b, which in turn is equivalent to Property 14.7c. While we do not recommend using the rewrite rules unless you are implementing a tool, looking at them is one way to gain a deeper understanding of nested clocks and of multiply-clocked properties in general.

NOTE: The rewrite rules are given for "basic" operators only. Rewrite rules for the remaining operators can be derived from the basic rewrite rules through equivalences such as those presented in Section B.4. For example, {`b[*3]`} is equivalent to {`b;b;b`}, thus rewrite rules for the form {`b[*n]`} are not given.

```
always ({a ; b@c2 ; c} |-> {d;e;f})@c1                    (14.7a)

always {a@c1 ; b@c2 ; c@c1} |-> {d@c1 ; e@c1 ; f@c1}      (14.7b)

always {!c1[*] ; a && c1;                                  (14.7c)
        !c2[*] ; b && c2;
        !c1[*] ; c && c1} |->
       {!c1[*] ; d && c1;
        !c1[*] ; e && c1;
        !c1[*] ; f && c1}
```

Fig. 14.7: Understanding clock nesting through the rewrite rules

14.4 A "behind the scenes" view of clocking

Up until now, we have explained that the clock operator in PSL has the effect of filtering out cycles where the clock does not hold. For singly-clocked properties, this is all you need to know. For multiply-clocked properties, it does not give the entire picture. We have given rules of thumb and examples, and mentioned the rewrite rules, but we have not explained directly the way that multiple clocks interact in a property. In order to use multiply-clocked properties correctly, it is necessary to understand a little bit more.

The complications arise from the fact that for multiply-clocked properties, it is necessary to filter out different cycles for each signal. How, though, do we move between the two sets of cycles? First we note that it is not as simple as merely looking at all cycles in which some clock holds. For instance, both the cycles of `@fell(clk1)` and those of `@fell(clk2)` are shaded in Trace 14.8(i).

We cannot understand Property 14.8a by viewing its unclocked version on the shaded cycles of Trace 14.8(i). The reason is that Trace 14.8(i) looks at signals `a` and `b` in all of the shaded cycles, whereas it should be looking at `a` in only some of them, and `b` in the others.

But why? Signal `a` can't change at cycles of `fell(clk2)`, and signal `b` can't change at cycles of `fell(clk1)`. So by the same reasoning that led us previously to conclude that we can use `@fell(clk1)` and `@fell(clk2)` in a design that works on the rising edges of `clk1` and `clk2`, we should be able to conclude that the "extra" cycle pairs (9,14) and (19,24), etc. should not bother us. While this happens to be true for our simple property, the same argument will not work in a property that counts cycles. For instance, if we want to say that if signal `cancel` is asserted for two consecutive cycles of `clk1`, then signal `flush_queue` should be asserted at the next cycle of `clk2`, we could code Assertion 14.9a. Assertion 14.9a holds on Trace 14.9(i), but the

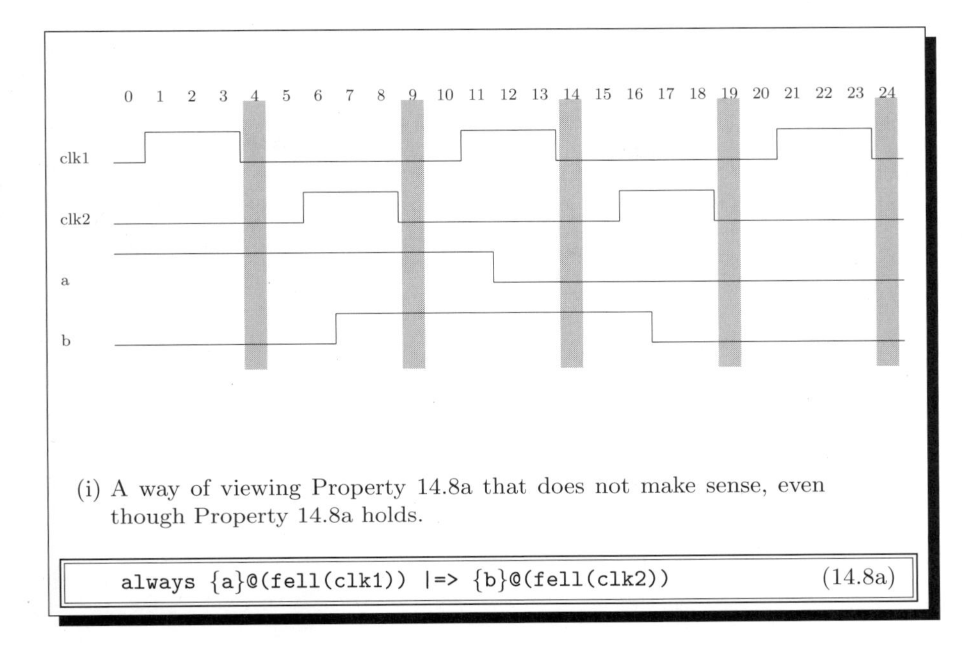

(i) A way of viewing Property 14.8a that does not make sense, even though Property 14.8a holds.

```
always {a}@(fell(clk1)) |=> {b}@(fell(clk2))     (14.8a)
```

Fig. 14.8: Choosing both clocks makes no sense

unclocked version does not hold on the trace formed from the shaded cycles of Trace 14.9(ii) because at (PSL) cycle 8 we have seen two consecutive shaded cycles in which `cancel` is asserted, yet signal `flush_queue` is not asserted on the next shaded cycle occurring at cycle 12.

So, we are now convinced that understanding multiply-clocked properties is not trivial. How then can they be understood? The answer is that a multiply-clocked property should be thought of as creating multiple sets of filtered cycles. For instance, the entire Assertion 14.9a is unclocked, so that creates the set of cycles shown in Trace 14.10(i). The left-hand side of the suffix implication creates the set of cycles shown in Trace 14.10(ii) and the right-hand side creates the set shown in Trace 14.10(iii). You "run" using the set of cycles relevant to the part of the property currently being examined. Thus, the entire Assertion 14.9a holds on Trace 14.9(i) if it holds on the trace composed of the shaded cycles of Trace 14.10(i) – which is all of them because the entire property is unclocked. In turn, that happens only if the sub-property `{cancel[*2]}@fell(clk1) |=> {flush_queue}@fell(clk2)` holds at every cycle of Trace 14.10(i). That happens only if, whenever a match of `cancel[*2]` holds on the trace composed of the shaded cycles of Trace 14.10(ii), then starting the cycle after the cycle where the match ended (shaded or unshaded, because the `|=>` operator is not clocked), we see a match of `flush_queue` on the trace composed of the shaded cycles of Trace 14.10(iii). For instance, in order for Assertion 14.9a to hold on Trace 14.9(i), `flush_queue` must hold on

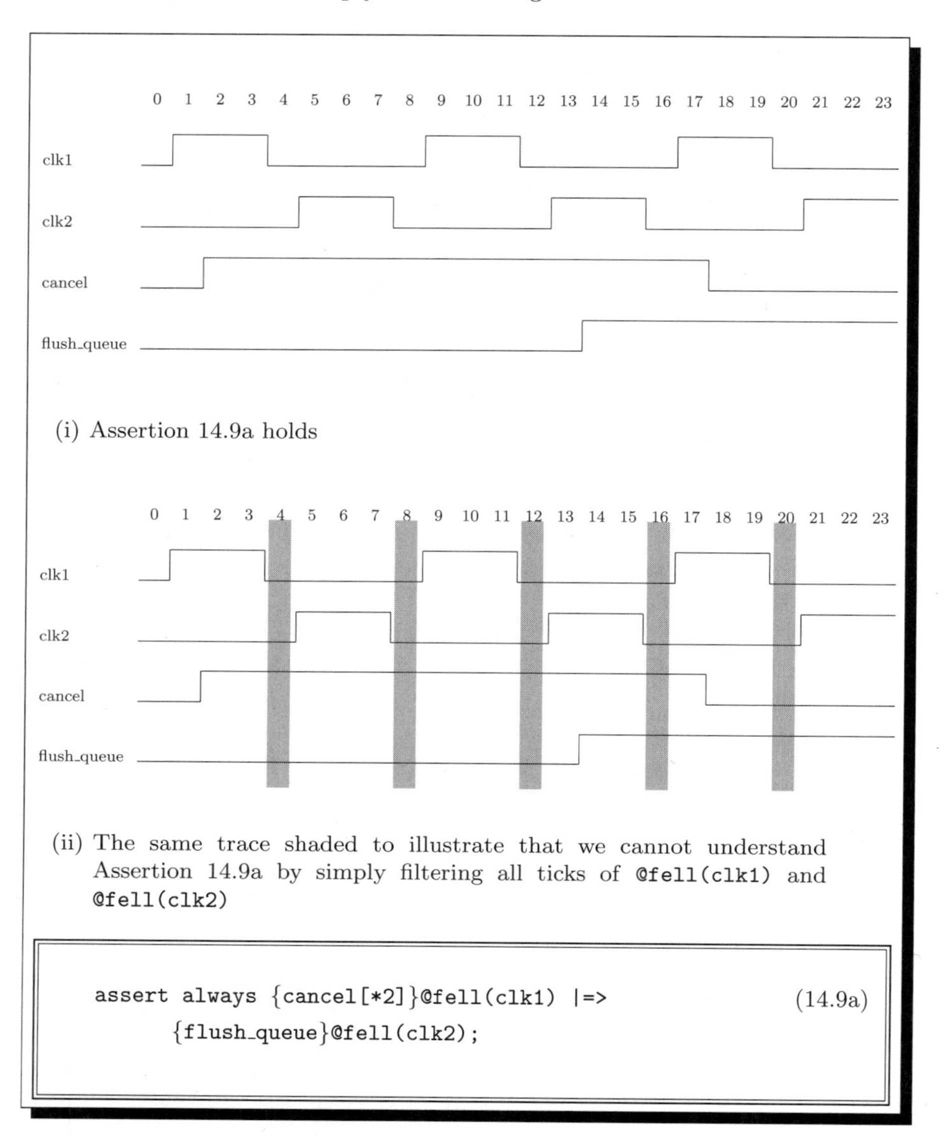


```
assert always {cancel[*2]}@fell(clk1) |=>                    (14.9a)
       {flush_queue}@fell(clk2);
```

Fig. 14.9: If counting cycles is important, we cannot choose both clocks

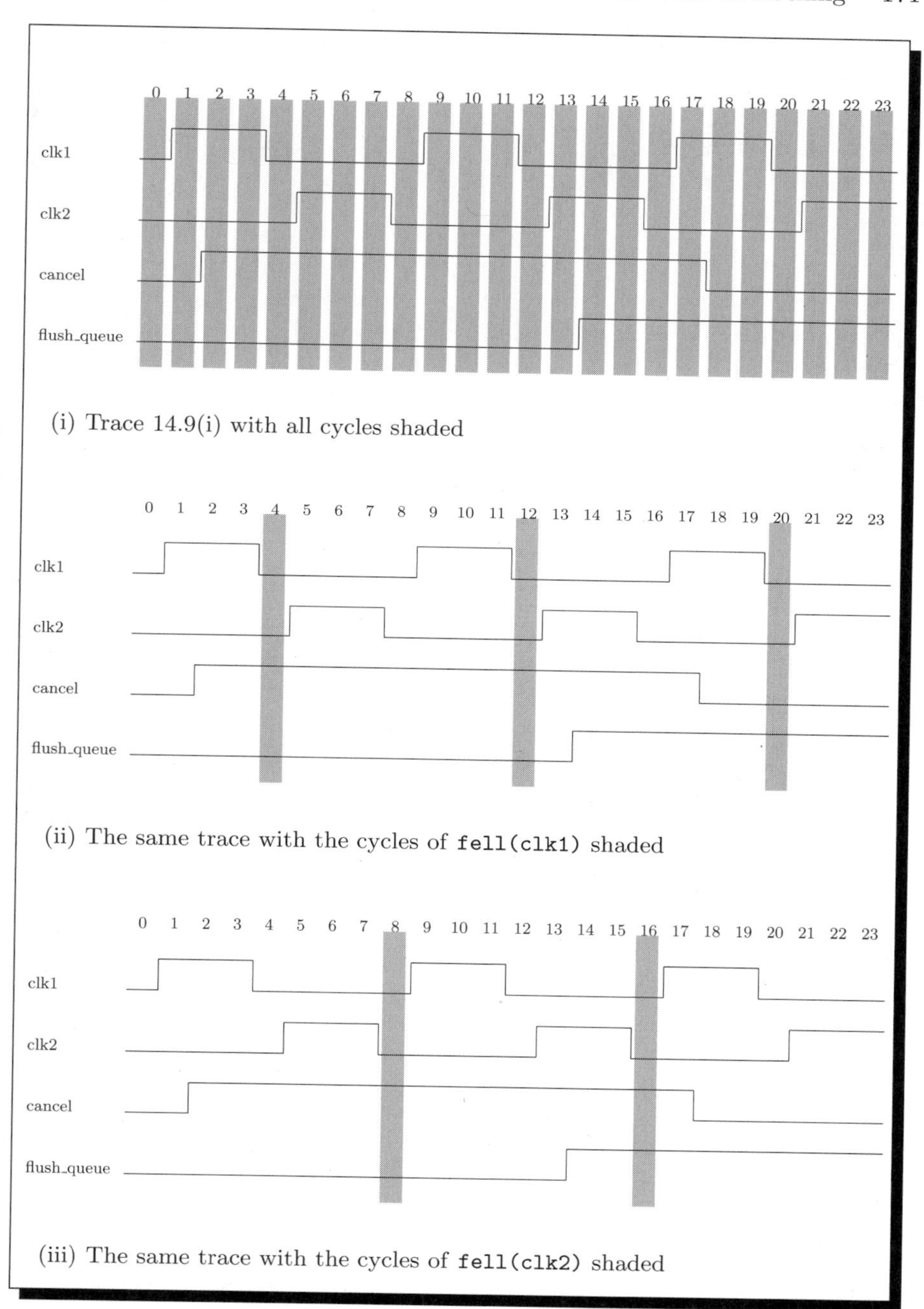

Fig. 14.10: A multiply-clocked property can be understood as creating multiple sets of filtered cycles

```
always ({a ; b@c2 ; c} |-> {d;e;f})@c1                    (14.11a)

always {!c1[*] ; a && c1;                                 (14.11b)
        !c2[*] ; b && c2;
        !c1[*] ; c && c1} |->
       {!c1[*] ; d && c1;
        !c1[*] ; e && c1;
        !c1[*] ; f && c1}
```

Fig. 14.11: Expressing a property without the clock operator is cumbersome

the shaded cycles of Trace 14.10(iii) starting at cycle 13, because a match of `cancel[*2]` completed at cycle 12 of Trace 14.10(ii).

Up until now, we've seen multiply-clocked properties in SERE style. In LTL style, multiply-clocked properties are possible, but are much more (visually) complicated, and much more complicated to understand. The reason is that LTL-style properties usually involve several temporal operators nested one within the other (whereas SERE style properties typically involve at most one degree of nesting). Thus, understanding a multiply-clocked LTL style property involves a lot of "jumping" from one set of filtered cycles to another. It is therefore strongly recommended to stick with SERE style when multiple clocks are required. If for some reason you feel you absolutely must use multiple clocks in LTL style, the rule of thumb is to clock the whole property on the clock of the driving signal or signals. The driven signal or signals should be clocked within the property using their own clock.

While using the clock operator in a multiply-clocked design – especially in LTL style – might take some time to master, formulating an equivalent property without the clock operator is much more cumbersome, as can be seen by comparing equivalent Properties 14.7a and 14.7c, repeated here as Properties 14.11a and 14.11b.

14.5 Clocks that are not well-behaved

Up until now, we have been assuming that multiple clocks are well-behaved, that is, that they are interleaved. For clocks that are not well-behaved, if we can nevertheless guarantee at most one tick of `clk1` (the driving clock) in between any two ticks of `clk2`, as shown in Trace 14.12(i), we can still clock as we did for well-behaved clocks. However, if we cannot guarantee at most one tick of `clk1` in between any two ticks of `clk2` – for instance, if our clocks

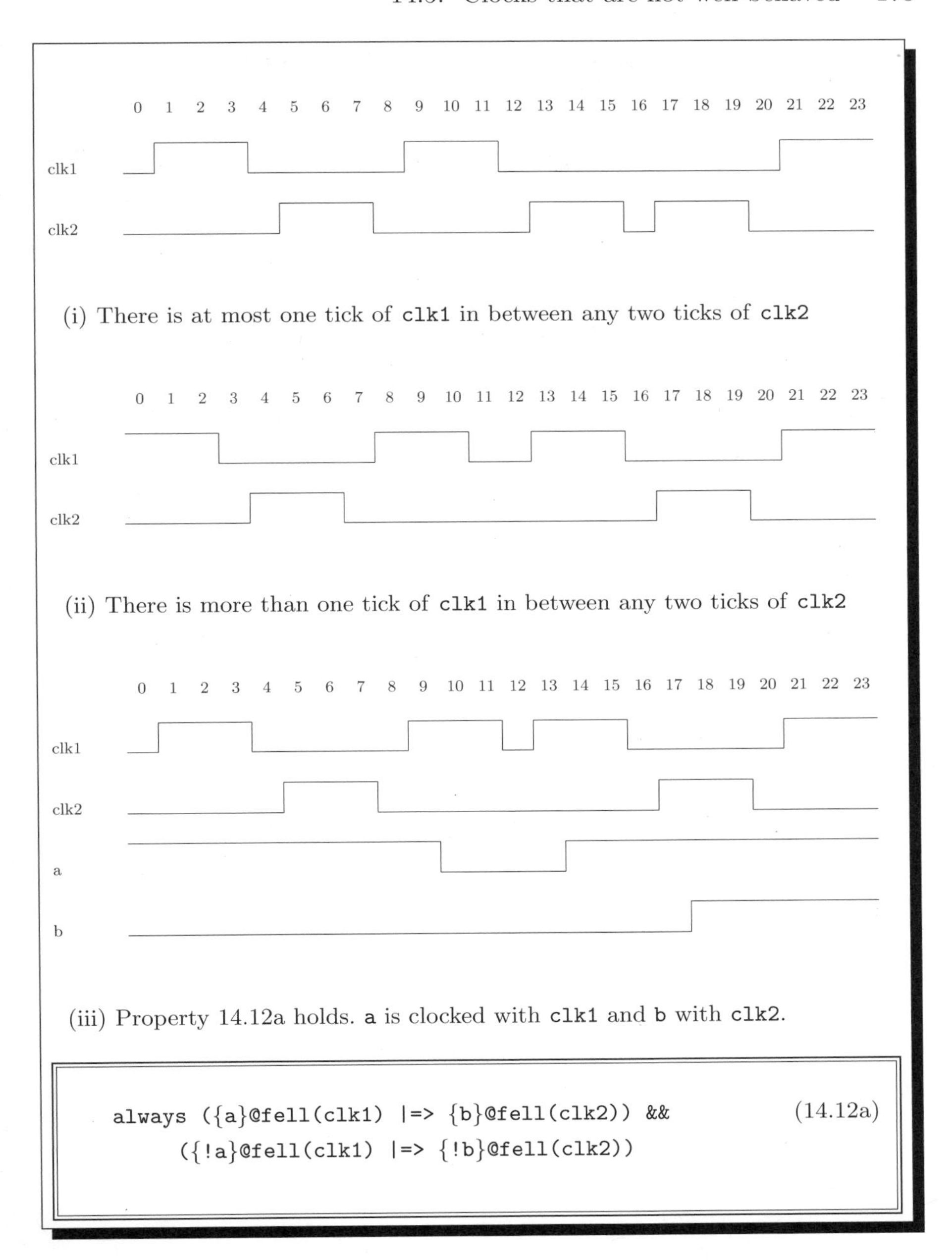

Fig. 14.12: Clocks that are not well-behaved

behave as shown in Trace 14.12(ii) – then the clock operator of PSL cannot be used to describe the situation shown in Figure 14.1.

To understand why, we need to write a complete description of the situation shown in Figure 14.1. That is, we need to mention not only what happens when `a` is asserted, but also what happens when it is deasserted, as shown in Property 14.12a. Property 14.12a says that if `a` holds on a tick of `clk1`, then `b` will hold on the next tick of `clk2`, and that if `a` does not hold on a tick of `clk1`, then `b` will not hold on the next tick of `clk2`. If we cannot guarantee at most one tick of `clk1` in between any two ticks of `clk2`, we might have the situation shown in Trace 14.12(iii). Signal `a` is clocked with `clk1`, which ticks more than once in between two ticks of `clk2`. On one tick of `clk1`, signal `a` is deasserted, and on the other, it is asserted. And both ticks have the same next tick of `clk2`. Thus Property 14.12a makes conflicting demands of Trace 14.12(iii), and thus does not hold on it.

Does this mean that PSL cannot be used to describe a multiply-clocked design in which the clocks are not well-behaved? No, it does not. Rather, what it means is that in such a design, Property 14.12a does not describe Figure 14.1. In such a design, `b` will latch the value of `a` on every second tick of `a`. Thus, we could code Property 14.13a instead.

```
always ({[*1]@fell(clk2) ; {[*1] ; a}@fell(clk1)}          (14.13a)
      |=>{b}@fell(clk2)) &&
      ({[*1]@fell(clk2) ; {[*1] ; !a}@fell(clk1)}
      |=>{!b}@fell(clk2))
```

Fig. 14.13: Clocks that are not well-behaved

Property 14.13a is quite complicated. Luckily, the onus is usually on the design and not the property. In other words, it is very unlikely that your intention is to describe the topology of the netlist as Property 14.13a does for Figure 14.1, and much more likely that the specification is something simple while it is up to the designer to meet the simple specification with a possibly complicated design.

A

Syntax Rule Summary

Below we present the syntax of PSL in Backus-Naur Form (BNF).

A.1 Conventions

The formal syntax described uses the following extended Backus-Naur Form (BNF).

a. The initial character of each word in a nonterminal is capitalized. For example:

   PSL_Statement

   A nonterminal is either a single word or multiple words separated by underscores. When a multiple word nonterminal containing underscores is referenced within the text (e.g., in a statement that describes the semantics of the corresponding syntax), the underscores are replaced with spaces.

b. Boldface words are used to denote reserved keywords, operators, and punctuation marks as a required part of the syntax. For example:

   vunit (;

c. The ::= operator separates the two parts of a BNF syntax definition. The syntax category appears to the left of this operator and the syntax description appears to the right of the operator. For example, item (d) shows three options for a *Vunit_Type*.

d. A vertical bar separates alternative items (use one only) unless it appears in boldface, in which case it stands for itself. For example:

Vunit_Type ::= **vunit** | **vprop** | **vmode**

e. Square brackets enclose optional items unless it appears in boldface, in which case it stands for itself. For example:

Sequence_Declaration ::=
    sequence Name [**(** Formal_Parameter_List **)**] DEF_SYM Sequence **;**

indicates that **(** *Formal_Parameter_List* **)** is an optional syntax item for *Sequence_Declaration*, whereas

| Sequence **[** ***** [Range] **]**

indicates that (the outer) square brackets are part of the syntax, while *Range* is optional.

f. Braces enclose a repeated item unless it appears in boldface, in which case it stands for itself. A repeated item may appear zero or more times; the repetitions occur from left to right as with an equivalent left-recursive rule. Thus, the following two rules are equivalent:

Formal_Parameter_List ::= Formal_Parameter { **;** Formal_Parameter }
Formal_Parameter_List ::=
    Formal_Parameter | Formal_Parameter_List **;** Formal_Parameter

g. A colon (:) in a production starts a line comment unless it appears in boldface, in which case it stands for itself.

h. If the name of any category starts with an italicized part, it is equivalent to the category name without the italicized part. The italicized part is intended to convey some semantic information. For example, *vunit*_Name is equivalent to Name.

i. Flavor macros, containing embedded underscores, are shown in uppercase. These reflect the various HDLs that can be used within the PSL syntax and show the definition for each HDL. The general format is the term Flavor Macro, then the actual *macro name*, followed by the = operator, and, finally, the definition for each of the HDLs. For example:

Flavor Macro RANGE_SYM =
    SystemVerilog: **:** / Verilog: **:** / VHDL: **to** / GDL: **/ ..**

shows the *range symbol* macro (RANGE_SYM). See for further details about *flavor macros.*

The main text uses *italicized* type when a term is being defined, and monospace font for examples and references to constants such as 0, 1, or x values.

A.2 Tokens

PSL syntax is defined in terms of primitive *tokens*, which are character sequences that act as distinct symbols in the language.

Each PSL keyword is a single token. Some keywords end in one or two non-alphabetic characters ('!' or '_' or both). Those characters are part of the keyword, not separate tokens.

Each of the following character sequences is also a token:

```
[     ]     (     )     {     }
,     ;     :     ..    =     :=
*     +     |->   |=>   <->   ->
[*    [+]   [->   [=
&&    &     ||    |     !
$     @     .     /
```

Finally, for a given flavor, the tokens of the corresponding HDL are tokens of PSL.

A.3 HDL dependencies

PSL depends upon the syntax and semantics of an underlying hardware description language. In particular, PSL syntax includes productions that refer to nonterminals in SystemVerilog, Verilog, VHDL, or GDL. PSL syntax also includes Flavor Macros that cause each flavor of PSL to match that of the underlying HDL for that flavor.

For SystemVerilog, the PSL syntax refers to the following nonterminals in the IEEE P1800 syntax:

- module_or_generate_item_declaration
- module_or_generate_item
- list_of_variable_identifiers
- identifier
- expression
- constant_expression

For Verilog, the PSL syntax refers to the following nonterminals in the IEEE Std 1364 syntax:

- module_or_generate_item_declaration
- module_or_generate_item
- list_of_variable_identifiers
- identifier
- expression
- constant_expression
- task_port_type

For VHDL, the PSL syntax refers to the following nonterminals in the IEEE Std 1076 syntax:

- block_declarative_item
- concurrent_statement
- design_unit
- identifier
- expression
- entity_aspect

For SystemC, the PSL syntax refers to the following nonterminals in the IEEE P1666 syntax:

- simple_type_specifier
- expression
- event_expression
- declaration
- statement
- identifier

For GDL, the PSL syntax refers to the following nonterminals in the GDL syntax:

- module_item_declaration
- module_item
- module_declaration
- identifier
- expression

A.3.1 Verilog extensions

For the Verilog flavor, PSL extends the forms of declaration that can be used in the modeling layer by defining two additional forms of type declaration.

Extended_Verilog_Declaration ::=
    *Verilog*_module_or_generate_item_declaration
  | Extended_Verilog_Type_Declaration

Extended_Verilog_Type_Declaration ::=
    Finite_Integer_Type_Declaration
  | Structure_Type_Declaration

Finite_Integer_Type_Declaration ::=
  integer Integer_Range list_of_variable_identifiers **;**

Structure_Type_Declaration ::=
  struct { Declaration_List **}** list_of_variable_identifiers **;**

Integer_Range ::=
  (constant_expression **:** constant_expression **)**

Declaration_List ::=
  HDL_Variable_or_Net_Declaration { HDL_Variable_or_Net_Declaration }

HDL_Variable_or_Net_Declaration ::=
    net_declaration
  | reg_declaration
  | integer_declaration

A.3.2 Flavor macros

Flavor Macro DEF_SYM =
  SystemVerilog: **=** / Verilog: **=** / VHDL: **is** / SystemC: **=** / GDL: **:=**

Flavor Macro RANGE_SYM =
  SystemVerilog: **:** / Verilog: **:** / VHDL: **to** / SystemC: **:** / GDL: **..**

Flavor Macro AND_OP =
  SystemVerilog: **&&** / Verilog: **&&** / VHDL: **and** / SystemC: **&&** / GDL: **&**

Flavor Macro OR_OP =
  SystemVerilog: || / Verilog: || / VHDL: **or** / SystemC: || / GDL: |

Flavor Macro NOT_OP =
SystemVerilog: **!** / Verilog: **!** / VHDL: **not** / SystemC: **!** / GDL: **!**

Flavor Macro MIN_VAL =
SystemVerilog: **0** / Verilog: **0** / VHDL: **0** / SystemC: **0** / GDL: *null*

Flavor Macro MAX_VAL =
SystemVerilog: **$** / Verilog: **inf** / VHDL: **inf** / SystemC: **inf** / GDL: *null*

Flavor Macro HDL_EXPR =
SystemVerilog: *SystemVerilog*_Expression
/ Verilog: *Verilog*_Expression
/ VHDL: *VHDL*_Expression
/ SystemC: *SystemC*_Expression
/ GDL: *GDL*_Expression

Flavor Macro HDL_CLOCK_EXPR =
SystemVerilog: *SystemVerilog*_Event_Expression
/ Verilog: *Verilog*_Event_Expression
/ VHDL: *VHDL*_Expression
/ SystemC: *SystemC*_Event_Expression
/ GDL: *GDL*_Expression

Flavor Macro HDL_UNIT =
SystemVerilog: *SystemVerilog*_module_declaration
/ Verilog: *Verilog*_module_declaration
/ VHDL: *VHDL*_design_unit
/ SystemC: *SystemC_class*_sc_module
/ GDL: *GDL*_module_declaration

Flavor Macro HDL_DECL =
SystemVerilog: *SystemVerilog*_module_or_generate_item_declaration
/ Verilog: Extended_Verilog_Declaration
/ VHDL: *VHDL*_block_declarative_item
/ SystemC: *SystemC*_declaration
/ GDL: *GDL*_module_item_declaration

Flavor Macro HDL_STMT =
SystemVerilog: *SystemVerilog*_module_or_generate_item
/ Verilog: *Verilog*_module_or_generate_item
/ VHDL: *VHDL*_concurrent_statement
/ SystemC: *SystemC*_statement
/ GDL: *GDL*_module_item

Flavor Macro HDL_VARIABLE_TYPE =
SystemVerilog : *SystemVerilog*_data_type
/ Verilog : *Verilog*_Variable_Type
/ VHDL : *VHDL*_subtype_indication
/ SystemC: *SystemC*_simple_type_specifier
/ GDL : *GDL*_variable_type

Flavor Macro HDL_RANGE =
VHDL: range_attribute_name

Flavor Macro LEFT_SYM =
SystemVerilog: [/ Verilog: [/ VHDL: (/ SystemC: (/ GDL: (

Flavor Macro RIGHT_SYM =
SystemVerilog:] / Verilog:] / VHDL:) / SystemC:) / GDL:)

A.4 Syntax productions

The rest of this appendix defines the PSL syntax.

A.4.1 Verification units

PSL_Specification ::=
{ Verification_Item }

Verification_Item ::=
HDL_UNIT | Verification_Unit

Verification_Unit ::=
Vunit_Type *PSL*_Identifier [(Hierarchical_HDL_Name)] **{**
{ Inherit_Spec }
{ Vunit_Item }
}

Vunit_Type ::=
vunit | **vprop** | **vmode**

Hierarchical_HDL_Name ::=
HDL_Module_NAME { Path_Separator *instance*_Name }

*instance*_Name ::=
HDL_or_PSL_Identifier

HDL_Module_Name ::=
  *HDL_Module*_Name [(*HDL_Module_Name*)]

Path_Separator ::=
  . | /

Inherit_Spec ::=
  inherit *vunit*_Name { , *vunit*_Name } ;

Vunit_Item ::=
    HDL_DECL
  | HDL_STMT
  | PSL_Declaration
  | PSL_Directive

A.4.2 PSL declarations

PSL_Declaration ::=
    Property_Declaration
  | Sequence_Declaration
  | Clock_Declaration

Property_Declaration ::=
  property *PSL*_Identifier [**(** Formal_Parameter_List **)**] DEF_SYM Property **;**

Formal_Parameter_List ::=
  Formal_Parameter { ; Formal_Parameter }

Formal_Parameter ::=
  Param_Spec *PSL*_Identifier { , *PSL*_Identifier }

Param_Spec ::=
    const
  | [**const**] Value_Parameter
  | **sequence**
  | **property**

Value_Parameter ::=
    HDL_Type
  | PSL_Type_Class

HDL_Type ::=

hdltype HDL_VARIABLE_TYPE

PSL_Type_Class ::= **boolean** | **bit** | **bitvector** | **numeric** | **string**

Sequence_Declaration ::=
  sequence *PSL*_Identifier [(Formal_Parameter_List)] DEF_SYM Sequence **;**

Clock_Declaration ::=
  default clock DEF_SYM Clock_Expression **;**

Clock_Expression ::=
    *boolean*_Name
  | *boolean*_Built_In_Function_Call
  | (Boolean)
  | (HDL_CLOCK_EXPR)

Actual_Parameter_List ::=
  Actual_Parameter { , Actual_Parameter }

Actual_Parameter ::=
  AnyType|Number | Boolean | Property | Sequence

A.4.3 PSL directives

PSL_Directive ::=
  [Label **:**] Verification_Directive

Label ::=
  *PSL*_Identifier

HDL_or_PSL_Identifier ::=
    *SystemVerilog*_Identifier
  | *Verilog*_Identifier
  | *VHDL*_Identifier
  | *SystemC*_Identifier
  | *GDL*_Identifier
  | *PSL*_Identifier

Verification_Directive ::=
    Assert_Directive
  | Assume_Directive
  | Assume_Guarantee_Directive
  | Restrict_Directive

| Restrict_Guarantee_Directive
| Cover_Directive
| Fairness_Statement

Assert_Directive ::=
assert Property [**report** String] ;

Assume_Directive ::=
assume Property **;**

Assume_Guarantee_Directive ::=
assume_guarantee Property [**report** String] ;

Restrict_Directive ::=
restrict Sequence ;

Restrict_Guarantee_Directive ::=
restrict_guarantee Sequence [**report** String] ;

Cover_Directive ::=
cover Sequence [**report** String] ;

Fairness_Statement ::=
fairness Boolean **;**
| **strong fairness** Boolean **,** Boolean **;**

A.4.4 PSL properties

Property ::=
Replicator Property
| FL_Property
| OBE_Property

Replicator ::=
forall Parameter_Definition **:**

Index_Range ::=
LEFT_SYM *finite*_Range RIGHT_SYM
| (HDL_RANGE)

Value_Set ::=
{ Value_Range { **,** Value_Range } **}**
| **boolean**

```
Value_Range ::=
    Value
  | finite_Range

Value ::=
    Boolean
  | Number

FL_Property ::=
    Boolean
  | ( FL_Property )
  | Sequence [ ! ]
  | FL_property_Name [ ( Actual_Parameter_List ) ]
  | FL_Property @ Clock_Expression
  | FL_Property abort Boolean
  | FL_Property async_abort Boolean
  | FL_Property sync_abort Boolean
  | Parameterized_Property
: Logical Operators :
  | NOT_OP FL_Property
  | FL_Property AND_OP FL_Property
  | FL_Property OR_OP FL_Property
  :
  | FL_Property -> FL_Property
  | FL_Property <-> FL_Property
: Primitive LTL Operators :
  | X FL_Property
  | X! FL_Property
  | F FL_Property
  | G FL_Property
  | [ FL_Property U FL_Property ]
  | [ FL_Property W FL_Property ]
: Simple Temporal Operators :
  | always FL_Property
  | never FL_Property
  | next FL_Property
  | next! FL_Property
  | eventually! FL_Property
  :
  | FL_Property until! FL_Property
  | FL_Property until FL_Property
  | FL_Property until!_ FL_Property
  | FL_Property until_ FL_Property
  :
```

| FL_Property **before!** FL_Property
| FL_Property **before** FL_Property
| FL_Property **before!_** FL_Property
| FL_Property **before_** FL_Property
: Extended Next (Event) Operators :
| **X** [Number] (FL_Property)
| **X!** [Number] (FL_Property)
| **next** [Number] (FL_Property)
| **next!** [Number] (FL_Property)
:
| **next_a** [*finite*_Range] (FL_Property)
| **next_a!** [*finite*_Range] (FL_Property)
| **next_e** [*finite*_Range] (FL_Property)
| **next_e!** [*finite*_Range] (FL_Property)
:
| **next_event!** (Boolean) (FL_Property)
| **next_event** (Boolean) (FL_Property)
| **next_event!** (Boolean) [*positive*_Number] (FL_Property)
| **next_event** (Boolean) [*positive*_Number] (FL_Property)
:
| **next_event_a!** (Boolean) [*finite_positive*_Range] (FL_Property)
| **next_event_a** (Boolean) [*finite_positive*_Range] (FL_Property)
| **next_event_e!** (Boolean) [*finite_positive*_Range] (FL_Property)
| **next_event_e** (Boolean) [*finite_positive*_Range] (FL_Property)
: Operators on SEREs :
| { SERE } (FL_Property)
| Sequence |-> FL_Property
| Sequence |=> FL_Property

A.4.5 Sequential Extended Regular Expressions (SEREs)

SERE ::=
Boolean
| Sequence
| SERE **;** SERE
| SERE **:** SERE
| Compound_SERE

Compound_SERE ::=
Repeated_SERE
| Braced_SERE
| Clocked_SERE
| Compound_SERE | Compound_SERE

| Compound_SERE **&** Compound_SERE
| Compound_SERE **&&** Compound_SERE
| Compound_SERE **within** Compound_SERE
| Parameterized_SERE

A.4.6 Parameterized Properties and SEREs

Parameterized_Property ::=
  for Parameters_Definition **:** And_Or_Property_OP (FL_Property)

Parameterized_SERE ::=
  for Parameters_Definition **:** And_Or_SERE_OP **{** SERE **}**

Parameters_Definition ::=
  Parameter_Definition { Parameter_Definition }

Parameter_Definition ::=
  PSL_Identifier [Index_Range] **in** Value_Set

And_OR_Property_OP ::=
   AND_OP
  | OR_OP

And_Or_SERE_Op :: =
  && | **&** | |

A.4.7 Sequences

Sequence ::=
   Sequence_Instance
  | Repeated_SERE
  | Braced_SERE
  | Clocked_SERE

Repeated_SERE ::=
   Boolean [* [Count]]
  | Sequence [* [Count]]
  | [* [Count]]
  | Boolean [+]
  | Sequence [+]
  | [+]

| Boolean [= Count]
| Boolean [-> [*positive*_Count]]

Braced_SERE ::=
{ SERE **}**

Sequence_Instance ::=
*sequence*_Name [**(** Actual_Parameter_List **)**]

Clocked_SERE ::=
Braced_SERE **@** Clock_Expression

Count ::=
Number
| Range

Range ::=
Low_Bound RANGE_SYM High_Bound

Low_Bound ::=
Number
| MIN_VAL

High_Bound ::=
Number
| MAX_VAL

A.4.8 Forms of expression

Any_Type ::=
HDL_or_PSL_Expression

Bit ::=
*bit*_HDL_or_PSL_Expression

Boolean ::=
*boolean*_HDL_or_PSL_Expression

BitVector ::=
*bitvector*_HDL_or_PSL_Expression

Number ::=
*numeric*_HDL_or_PSL_Expression

String ::=
*string*_HDL_or_PSL_Expression

HDL_or_PSL_Expression ::=
HDL_Expression
| PSL_Expression
| Built_In_Function_Call
| Union_Expression

HDL_Expression ::=
HDL_EXPR

PSL_Expression ::=
Boolean **->** Boolean
| Boolean <–> Boolean

Built_In_Function_Call ::=
prev (Any_Type [, Number [, Clock_Expression]])
| **next** (Any_Type)
| **stable** (Any_Type [, Clock_Expression])
| **rose** (Bit [, Clock_Expression])
| **fell** (Bit [, Clock_Expression])
| **ended** (Sequence [, Clock_Expression])
| **isunknown** (BitVector)
| **countones** (BitVector)
| **onehot** (BitVector)
| **onehot0** (BitVector)
| **nondet** (Value_List)
| **nondet_vector** (Number, Value_List)

Union_Expression ::=
Any_Type **union** Any_Type

A.4.9 Optional Branching Extension

OBE_Property ::=
Boolean
| (OBE_Property)
| *OBE_property*_Name [(Actual_Parameter_List)]
: Logical Operators :
| NOT_OP OBE_Property
| OBE_Property AND_OP OBE_Property

```
    | OBE_Property OR_OP OBE_Property
    | OBE_Property -> OBE_Property
    | OBE_Property <–> OBE_Property
: Universal Operators :
    | AX OBE_Property
    | AG OBE_Property
    | AF OBE_Property
    | A [ OBE_Property U OBE_Property ]
:Existential Operators :
    | EX OBE_Property
    | EG OBE_Property
    | EF OBE_Property
    | E [ OBE_Property U OBE_Property ]
```

B

Formal Syntax and Semantics

This appendix formally describes the syntax and semantics of the temporal layer.

B.1 Typed-text representation of symbols

Table B.1 shows the mapping of various symbols used in this definition to the corresponding typed-text PSL representation, in the different flavors.

Table B.1: Typed-text symbols in the SystemVerilog, Verilog, VHDL, SystemC and GDL flavors

<table>
<tr><th></th><th>SystemVerilog</th><th>Verilog</th><th>VHDL</th><th>SystemC</th><th>GDL</th></tr>
<tr><td>$\mapsto$</td><td>|-></td><td>|-></td><td>|-></td><td>|-></td><td>|-></td></tr>
<tr><td>$\Mapsto$</td><td>|=></td><td>|=></td><td>|=></td><td>|=></td><td>|=></td></tr>
<tr><td>$\rightarrow$</td><td>-></td><td>-></td><td>-></td><td>-></td><td>-></td></tr>
<tr><td>$\leftrightarrow$</td><td><-></td><td><-></td><td><-></td><td><-></td><td><-></td></tr>
<tr><td>$\neg$</td><td>!</td><td>!</td><td>not</td><td>!</td><td>!</td></tr>
<tr><td>$\wedge$</td><td>&&</td><td>&&</td><td>and</td><td>&&</td><td>&</td></tr>
<tr><td>$\vee$</td><td>||</td><td>||</td><td>or</td><td>||</td><td>|</td></tr>
<tr><td>$..$</td><td>:</td><td>:</td><td>to</td><td>:</td><td>..</td></tr>
<tr><td>$\langle\ \rangle$</td><td>[ ]</td><td>[ ]</td><td>( )</td><td>( )</td><td>( )</td></tr>
</table>

NOTE –
For reasons of simplicity, the syntax given herein is more flexible than the one defined by the extended BNF (given in Appendix A). That is, some of the expressions which are legal here are not legal under the BNF grammar. Users should use the stricter syntax, as defined by the BNF grammar in Appendix A.

B.2 Syntax

The logic PSL is defined with respect to a non-empty set of atomic propositions P and a given set of boolean expressions B over P. We assume two designated boolean expression *true* and *false* belong to B.

Definition 1 (Sequential Extended Regular Expressions (SEREs))

1. *Every boolean expression* $b \in B$ *is a SERE.*
2. *If* r, r_1, *and* r_2 *are SEREs, and* c *is a boolean expression, then the following are SEREs:*
   - $\{r\}$
   - $r_1 \; ; \; r_2$
   - $r_1 : r_2$
   - $r_1 \mid r_2$
   - $r_1 \;\&\&\; r_2$
   - $[*0]$
   - $r[*]$
   - $r@c$

Definition 2 (FL formulas)

1. *If* b *is a boolean expression, then both* b *and* $b!$ *are FL formulas.*
2. *If* φ *and* ψ *are FL formulas,* r, r_1, r_2 *are SEREs, and* b *a boolean expression, then the following are FL formulas:*
   - (φ)
   - $\neg\varphi$
   - $\varphi \wedge \psi$
   - $r \mapsto \varphi$
   - $r!$
   - r
   - $X! \; \varphi$
   - $[\varphi \; U \; \psi]$
   - $\varphi@b$
   - φ async_abort b
   - φ sync_abort b

NOTE –
We define formal semantics for both strong and weak booleans[20]. However, strong booleans are not accessible to the user.

Definition 3 (OBE Formulas)

1. *Every boolean expression is an OBE formula.*
2. *If* f, f_1, *and* f_2 *are OBE formulas, then so are the following:*
   a) (f)
   b) $\neg f$
   c) $f_1 \wedge f_2$
   d) EXf
   e) $E[f_1 \; U \; f_2]$
   f) EGf

Additional OBE operators are derived from these as follows:

- $f_1 \vee f_2 = \neg(\neg f_1 \wedge \neg f_2)$
- $f_1 \rightarrow f_2 = \neg f_1 \vee f_2$
- $f_1 \leftrightarrow f_2 = (f_1 \rightarrow f_2) \wedge (f_2 \rightarrow f_1)$
- $EFf = E[true \; U \; f]$
- $AXf = \neg EX \neg f$
- $A[f_1 \; U \; f_2] = \neg(E[\neg f_2 \; U \; (\neg f_1 \wedge \neg f_2)] \vee EG \neg f_2)$
- $AGf = \neg E[true \; U \; \neg f]$

- $AFf = A[true\ U\ f]$

Definition 4 (PSL Formulas)

1. *Every FL formula is a PSL formula.*
2. *Every OBE formula is a PSL formula.*

In Section B.4, we show additional operators which provide syntactic sugaring to the ones above.

B.3 Semantics

The semantics of PSL formulas are defined with respect to a *model*. A model is a quintuple (S, S_0, R, P, L), where S is a finite set of states, $S_0 \subseteq S$ is a set of initial states, $R \subseteq S \times S$ is the transition relation, P is a non-empty set of atomic propositions, and L is the valuation, a function $L : S \longrightarrow 2^P$, mapping each state with a set of atomic propositions valid in that state.

A *path* π is a finite (or infinite) sequence of states $\pi = (\pi_0, \pi_1, \pi_2, \cdots, \pi_n)$ (or $\pi = (\pi_0, \pi_1, \pi_2, \cdots)$). A *computation path* π of a model M is a finite (or infinite) path π such that for every $i < n$, $R(\pi_i, \pi_{i+1})$ and for no s, $R(\pi_n, s)$ (or such that for every i, $R(\pi_i, \pi_{i+1})$). Given a finite (or infinite) path π, we define $\hat{L}$, an extension of the valuation function L from states to paths as follows: $\hat{L}(\pi) = L(\pi_0)L(\pi_1)\ldots L(\pi_n)$ (or $\hat{L}(\pi) = L(\pi_0)L(\pi_1)\ldots$). Thus we have a mapping from states in M to letters of 2^P, and from finite (or infinite) sequences of states in M to finite (or infinite) words over 2^P.

B.3.1 Semantics of FL formulas

The semantics of FL formulas is interpreted over finite and infinite words over $\Sigma = 2^P \cup \{\top, \bot\}$. Let φ be an FL formula, w a word over Σ and M a model. The notation $w \models \varphi$ means that the FL formula φ holds over the word w. The notation $M \models \varphi$ means that for all π such that π is computation path of M, $\hat{L}(\pi) \models \varphi$.

We denote a letter from Σ by ℓ and an empty, finite, or infinite word from Σ by u, v, or w (possibly with subscripts). We denote the length of word v as $|v|$. A finite non-empty word $v = (\ell_0\ell_1\ell_2\cdots\ell_n)$ has length $n+1$, the (finite) empty word $v = \epsilon$ has length 0, and an infinite word has length ∞. We use i, j, and k to denote non-negative integers. We denote the ith letter of v by v^{i-1} (since counting of letters starts at zero). We denote by $v^{i..}$ the suffix of v starting at v^i. That is, for every $i < |v|$, $v^{i..} = v^iv^{i+1}\cdots v^n$ or $v^{i..} = v^iv^{i+1}\cdots$. We denote by $v^{i..j}$ the finite sequence of letters starting from v^i and ending in v^j. That is, for $j \geq i$, $v^{i..j} = v^iv^{i+1}\cdots v^j$ and for $j < i$, $v^{i..j} = \epsilon$. We use ℓ^ω to denote an infinite-length word, each letter of which is ℓ.

We use $\overline{v}$ to denote the word obtained by replacing every $\top$ with a $\bot$ and vice versa. We call $\overline{v}$ the *dual* of v.

The semantics of FL *formulas* over *words* is defined inductively, using as the base case the semantics of *boolean expressions* over *letters* in Σ. The semantics of boolean expression is assumed to be given as a relation $\Vdash\ \subseteq \Sigma \times B$ relating letters in Σ with boolean expressions in B. If $(\ell, b) \in \Vdash$ we say that the letter ℓ *satisfies* the boolean expression b and denote it $\ell \Vdash b$. We assume the two special letters $\top$ and $\bot$ behave as follows: for every boolean expression b, $\top \Vdash b$ and $\bot \not\Vdash b$. We assume that otherwise the boolean relation $\Vdash$ behaves in the usual manner. In particular, that for every letter $\ell \in 2^P$, atomic proposition $p \in P$ and boolean expressions $b, b_1, b_2 \in B$ (i) $\ell \Vdash p$ iff $p \in \ell$, (ii) $\ell \Vdash \neg b$ iff $\ell \not\Vdash b$, and (iii) $\ell \Vdash$ *true* and $\ell \not\Vdash$ *false*. Finally, we assume that for every letter $\ell \in \Sigma$, $\ell \Vdash b_1 \wedge b_2$ iff $\ell \Vdash b_1$ and $\ell \Vdash b_2$.

Unclocked Semantics

Semantics of unclocked SEREs

Unclocked SEREs are defined over finite words over the alphabet Σ. The notation $v \models\!\!\equiv r$, where r is a SERE and v a finite word means that v *models tightly* r. The semantics of unclocked SEREs are defined as follows, where b denotes a boolean expression, and r, r_1, and r_2 denote unclocked SEREs:

1. $v \equiv\!\!\models \{r\} \Longleftrightarrow v \equiv\!\!\models r$
2. $v \equiv\!\!\models b \Longleftrightarrow |v| = 1$ and $v^0 \Vdash b$
3. $v \equiv\!\!\models r_1 \ ; \ r_2 \Longleftrightarrow \exists v_1, v_2$ s.t. $v = v_1 v_2$, $v_1 \equiv\!\!\models r_1$, and $v_2 \equiv\!\!\models r_2$
4. $v \equiv\!\!\models r_1 : r_2 \Longleftrightarrow \exists v_1, v_2$, and ℓ s.t. $v = v_1 \ell v_2$, $v_1 \ell \equiv\!\!\models r_1$, and $\ell v_2 \equiv\!\!\models r_2$
5. $v \equiv\!\!\models r_1 \mid r_2 \Longleftrightarrow v \equiv\!\!\models r_1$ or $v \equiv\!\!\models r_2$
6. $v \equiv\!\!\models r_1 \;\&\&\; r_2 \Longleftrightarrow v \equiv\!\!\models r_1$ and $v \equiv\!\!\models r_2$
7. $v \equiv\!\!\models [*0] \Longleftrightarrow v = \epsilon$
8. $v \equiv\!\!\models r[*] \Longleftrightarrow$ either $v \equiv\!\!\models [*0]$
   or $\exists v_1, v_2$ s.t. $v_1 \neq \epsilon$, $v = v_1 v_2$, $v_1 \equiv\!\!\models r$ and $v_2 \equiv\!\!\models r[*]$

Semantics of unclocked FL

We refer to a formula of FL with no @ operator as an *unclocked formula*. Let v be a finite or infinite word, b be a boolean expression, r, r_1, r_2 unclocked SEREs, and φ, ψ unclocked FL formulas. We use $\models$ to define the semantics of unclocked FL formulas: If $v \models \varphi$ we say that v *models* (or *satisfies*) φ.

1. $v \models (\varphi) \Longleftrightarrow v \models \varphi$
2. $v \models \neg\varphi \Longleftrightarrow \overline{v} \not\models \varphi$
3. $v \models \varphi \wedge \psi \Longleftrightarrow v \models \varphi$ and $v \models \psi$
4. $v \models b! \Longleftrightarrow |v| > 0$ and $v^0 \Vdash b$

5. $v \models b \iff |v| = 0$ or $v^0 \Vdash b$
6. $v \models r \mapsto \varphi \iff \forall j < |v|$ s.t. $\overline{v}^{0..j} \models r$, $v^{j..} \models \varphi$
7. $v \models r! \iff \exists j < |v|$ s.t. $v^{0..j} \models r$
8. $v \models r \iff \forall j < |v|$, $v^{0..j}\top^\omega \models r!$
9. $v \models X!\, \varphi \iff |v| > 1$ and $v^{1..} \models \varphi$
10. $v \models [\varphi U \psi] \iff \exists k < |v|$ s.t. $v^{k..} \models \psi$, and $\forall j < k$, $v^{j..} \models \varphi$
11. $v \models \varphi$ async_abort $b \iff$ either $v \models \varphi$
    or $\exists j < |v|$ s.t. $v^j \Vdash b$ and $v^{0..j-1}\top^\omega \models \varphi$
12. $v \models \varphi$ sync_abort $b \iff$ either $v \models \varphi$
    or $\exists j < |v|$ s.t. $v^j \Vdash b$ and $v^{0..j-1}\top^\omega \models \varphi$

NOTES –

1. The semantics given here for the LTL operators and the async_abort operator is equivalent to the truncated semantics given in [18] which is interpreted over 2^P rather than over $2^P \cup \{\top, \bot\}$. Using $\models_\bullet$ for the semantics in [18], the following proposition states the equivalence: Let w be a finite word over 2^P, and let φ be a formula of LTL^{trunc}. Then, as shown in [19], the three following equivalences hold:

$$w \models^{-}_{\bullet} \varphi \iff w\top^\omega \models \varphi$$
$$w \models_{\bullet} \varphi \iff w \models \varphi$$
$$w \models^{+}_{\bullet} \varphi \iff w\bot^\omega \models \varphi$$

2. Using $\models_\bullet$ as in the note 1 above, we use *holds strongly* for $\models^{+}_{\bullet}$, *holds* for $\models_\bullet$, and *holds weakly* for $\models^{-}_{\bullet}$. The remaining terminology of Section 11.1 is formally defined as follows:
   - φ is *pending* on word w iff $w \models^{-}_{\bullet} \varphi$ and $w \not\models_\bullet \varphi$
   - φ *fails* on word w iff $w \not\models^{-}_{\bullet} \varphi$
3. There is a subtle difference between boolean negation and formula negation. For instance, consider the formula $\neg b$. If $\neg$ is boolean negation, then $\neg b$ holds on an empty path. If $\neg$ is formula negation, then $\neg b$ does not hold on an empty path. Rather than introduce distinct operators for boolean and formula negation, we instead adopt the convention that negation applied to a boolean expression is boolean negation. This does not restrict expressivity, as formula negation of b can be expressed as $(\neg b)!$.

Clocked Semantics

We say that finite word v *is a clock tick of* c iff $|v| > 0$ and $v^{|v|-1} \Vdash c$ and for every natural number $i < |v| - 1$, $v^i \Vdash \neg c$.

Semantics of clocked SEREs

Clocked SEREs are defined over finite words from the alphabet Σ and a boolean expression that serves as the clock context. The notation $v \stackrel{c}{|\!\equiv} r$, where r is a SERE, c is a boolean expression and v a finite word, means that v *models tightly* r *in context of clock* c. The semantics of clocked SEREs are defined as follows, where b, c, and c_1 denote boolean expressions, and r, r_1, and r_2 denote clocked SEREs:

1. $v \stackrel{c}{|\!\equiv} \{r\} \iff v \stackrel{c}{|\!\equiv} r$
2. $v \stackrel{c}{|\!\equiv} b \iff v$ is a clock tick of c and $v^{|v|-1} \Vdash b$
3. $v \stackrel{c}{|\!\equiv} r_1 \; ; \; r_2 \iff \exists v_1, v_2$ s.t. $v = v_1 v_2$, $v_1 \stackrel{c}{|\!\equiv} r_1$, and $v_2 \stackrel{c}{|\!\equiv} r_2$
4. $v \stackrel{c}{|\!\equiv} r_1 : r_2 \iff \exists v_1, v_2$, and ℓ s.t. $v = v_1 \ell v_2$, $v_1 \ell \stackrel{c}{|\!\equiv} r_1$, and $\ell v_2 \stackrel{c}{|\!\equiv} r_2$
5. $v \stackrel{c}{|\!\equiv} r_1 \mid r_2 \iff v \stackrel{c}{|\!\equiv} r_1$ or $v \stackrel{c}{|\!\equiv} r_2$
6. $v \stackrel{c}{|\!\equiv} r_1 \; \&\& \; r_2 \iff v \stackrel{c}{|\!\equiv} r_1$ and $v \stackrel{c}{|\!\equiv} r_2$
7. $v \stackrel{c}{|\!\equiv} [*0] \iff v = \epsilon$
8. $v \stackrel{c}{|\!\equiv} r[*] \iff$ either $v \stackrel{c}{|\!\equiv} [*0]$
   or $\exists v_1, v_2$ s.t. $v_1 \neq \epsilon$, $v = v_1 v_2$, $v_1 \stackrel{c}{|\!\equiv} r$ and $v_2 \stackrel{c}{|\!\equiv} r[*]$
9. $v \stackrel{c}{|\!\equiv} r@c_1 \iff v \stackrel{c1}{|\!\equiv} r$

Semantics of clocked FL

The semantics of (clocked) FL formulas is defined with respect to finite/infinite words over Σ and a boolean expression c which serves as the clock context. Let v be a finite or infinite word, b, c, c_1 boolean expressions, r, r_1, r_2 SEREs, and φ, ψ FL formulas. We use $\stackrel{c}{\models}$ to define the semantics of FL formulas. If $v \stackrel{c}{\models} \varphi$ we say that v *models* (or *satisfies*) φ *in the context of clock* c.

1. $v \stackrel{c}{\models} (\varphi) \iff v \stackrel{c}{\models} \varphi$
2. $v \stackrel{c}{\models} \neg\varphi \iff \overline{v} \stackrel{c}{\not\models} \varphi$
3. $v \stackrel{c}{\models} \varphi \wedge \psi \iff v \stackrel{c}{\models} \varphi$ and $v \stackrel{c}{\models} \psi$
4. $v \stackrel{c}{\models} b! \iff \exists j < |v|$ s.t. $v^{0..j}$ is a clock tick of c and $v^j \Vdash b$
5. $v \stackrel{c}{\models} b \iff \forall j < |v|$ s.t. $\overline{v}^{0..j}$ is a clock tick of c, $v^j \Vdash b$
6. $v \stackrel{c}{\models} r \mapsto \varphi \iff \forall j < |v|$ s.t. $\overline{v}^{0..j} \stackrel{c}{|\!\equiv} r$, $v^{j..} \stackrel{c}{\models} \varphi$
7. $v \stackrel{c}{\models} r! \iff \exists j < |v|$ s.t. $v^{0..j} \stackrel{c}{|\!\equiv} r$
8. $v \stackrel{c}{\models} r \iff \forall j < |v|$, $v^{0..j}\top^{\omega} \stackrel{c}{\models} r!$
9. $v \stackrel{c}{\models} X!\, f \iff \exists j < k < |v|$ s.t. $v^{0..j}$ and $v^{j+1..k}$ are clock ticks of c and
   $v^{k..} \stackrel{c}{\models} f$
10. $v \stackrel{c}{\models} [\varphi U \psi] \iff \exists k < |v|$ s.t. $v^k \Vdash c$, $v^{k..} \stackrel{c}{\models} \psi$, and
    $\forall j < k$ s.t. $v^j \Vdash c$, $v^{j..} \stackrel{c}{\models} \varphi$

11. $v \models^{c} \varphi @ c_1 \iff v \models^{c_1} \varphi$
12. $v \models^{c} \varphi$ $\mathsf{async\_abort}$ $b \iff$ either $v \models^{c} \varphi$
    or $\exists j < |v|$ s.t. $v^j \Vdash b$ and $v^{0..j-1}\top^\omega \models^{c} \varphi$
13. $v \models^{c} \varphi$ $\mathsf{sync\_abort}$ $b \iff$ either $v \models^{c} \varphi$ or
    or $\exists j < |v|$ s.t. $v^j \Vdash b \wedge c$ and $v^{0..j-1}\top^\omega \models^{c} \varphi$

NOTE –
The clocked semantics for the LTL subset follows the clocks paper [20], with the exception that strength is applied at the boolean level rather than at the propositional level.

B.3.2 Semantics of OBE formulas

The semantics of OBE formulas are defined over states in the *model*, rather than finite or infinite words. Let f be an OBE formula, $M = (S, S_0, R, P, L)$ a model and $s \in S$ a state of the model. The notation $M, s \models f$ means that f holds in state s of model M. The notation $M \models f$ is equivalent to $\forall s \in S_0 : M, s \models f$. In other words, f is valid for every initial state of M.

The semantics of OBE formulas are defined inductively, using as the base case the semantics of *boolean expressions* over *letters* in 2^P. The semantics of boolean expression is assumed to be given as a relation $\Vdash\ \subseteq 2^P \times B$ relating letters in 2^P with boolean expressions in B. If $(\ell, b) \in\ \Vdash$ we say that the letter ℓ *satisfies* the boolean expression b and denote it $\ell \Vdash b$. We assume that the boolean relation $\Vdash$ behaves in the usual manner. In particular, that for every letter $\ell \in 2^P$, atomic proposition $p \in P$ and boolean expressions $b, b_1, b_2 \in B$ (i) $\ell \Vdash p$ iff $p \in \ell$, (ii) $\ell \Vdash \neg b$ iff $\ell \not\Vdash b$, (iii) $\ell \Vdash b_1 \wedge b_2$ iff $\ell \Vdash b_1$ and $\ell \Vdash b_2$, and (iv) $\ell \Vdash$ *true* and $\ell \not\Vdash$ *false*.

The semantics of an OBE formula are those of standard CTL. The semantics are defined as follows, where b denotes a boolean expression and f, f_1, and f_2 denote OBE formulas:

1. $M, s \models b \iff L(s) \Vdash b$
2. $M, s \models (f) \iff M, s \models f$
3. $M, s \models \neg f \iff M, s \not\models f$
4. $M, s \models f_1 \wedge f_2 \iff M, s \models f_1$ and $M, s \models f_2$
5. $M, s \models EX\ f \iff$ there exists a computation path π of M such that $|\pi| > 1$, $\pi_0 = s$, and $M, \pi_1 \models f$
6. $M, s \models E[f_1\ U\ f_2] \iff$ there exists a computation path π of M such that $\pi_0 = s$ and there exists $k < |\pi|$ such that $M, \pi_k \models f_2$ and for every j such that $j < k$: $M, \pi_j \models f_1$
7. $M, s \models EG\ f \iff$ there exists a computation path π of M such that $\pi_0 = s$ and for every j such that $0 \leq j < |\pi|$: $M, \pi_j \models f$

B.4 Syntactic Sugaring

The remainder of the temporal layer is syntactic sugar. In other words, it does not add expressive power, and every piece of syntactic sugar can be defined in terms of the basic FL operators presented above. The syntactic sugar is defined below.

NOTE –
The definitions given here do not necessarily represent the most efficient implementation. In some cases, there is an equivalent syntactic sugaring, or a direct implementation, that is more efficient.

B.4.1 Additional SERE operators

Let i, j, k, and l be integer constants such that $i \geq 0, j \geq i, k \geq 1, l \geq k$. Then, additional SERE operators can be viewed as abbreviations of the basic SERE operators defined above, as follows, where b denotes a boolean expression, and r denotes a SERE:

- $r[+] \stackrel{\text{def}}{=} r; r[*]$
- $r[*0] \stackrel{\text{def}}{=} [*0]$
- $r[*k] \stackrel{\text{def}}{=} \overbrace{r; r; ...; r}^{k\ times}$
- $r[*i..j] \stackrel{\text{def}}{=} r[*i] \mid ... \mid r[*j]$
- $r[*i..] \stackrel{\text{def}}{=} r[*i]; r[*]$
- $r[*..i] \stackrel{\text{def}}{=} r[*0] \mid ... \mid r[*i]$
- $r[*..] \stackrel{\text{def}}{=} r[*0..]$
- $[+] \stackrel{\text{def}}{=} true[+]$
- $[*] \stackrel{\text{def}}{=} true[*]$
- $[*i] \stackrel{\text{def}}{=} true[*i]$
- $[*i..j] \stackrel{\text{def}}{=} true[*i..j]$
- $[*i..] \stackrel{\text{def}}{=} true[*i..]$
- $[*..i] \stackrel{\text{def}}{=} true[*..i]$
- $[*..] \stackrel{\text{def}}{=} true[*..]$
- $b[=i] \stackrel{\text{def}}{=} \{\neg b[*]; b\}[*i]; \neg b[*]$
- $b[=i..j] \stackrel{\text{def}}{=} b[=i] \mid ... \mid b[=j]$
- $b[=i..] \stackrel{\text{def}}{=} b[=i]; [*]$
- $b[=..i] \stackrel{\text{def}}{=} b[=0] \mid ... \mid b[=i]$
- $b[=..] \stackrel{\text{def}}{=} b[=0..]$
- $b[\rightarrow] \stackrel{\text{def}}{=} \neg b[*]; b$
- $b[\rightarrow k] \stackrel{\text{def}}{=} \{\neg b[*]; b\}[*k]$

- $b[\to k..l] \stackrel{\text{def}}{=} b[\to k] \mid ... \mid b[\to l]$
- $b[\to k..] \stackrel{\text{def}}{=} b[\to k] \mid \{b[\to k]; [*]; b\}$
- $b[\to ..k] \stackrel{\text{def}}{=} b[\to 1] \mid ... \mid b[\to k]$
- $b[\to ..] \stackrel{\text{def}}{=} b[\to 1..]$
- $r_1 \,\&\, r_2 \stackrel{\text{def}}{=} \{\{r_1\} \,\&\&\, \{r_2; true[*]\}\} \mid \{\{r_1; true[*]\} \,\&\&\, \{r_2\}\}$
- $r_1 \; within \; r_2 \stackrel{\text{def}}{=} \{[*];\ r_1;\ [*]\} \,\&\&\, \{r_2\}$

B.4.2 Additional FL operators

Let i, j, k and l be integers such that $i \geq 0$, $j \geq i$, $k > 0$ and $l \geq k$. Then, additional operators can be viewed as abbreviations of the basic operators defined above, as follows, where b denotes a boolean expression, r, r_1, and r_2 denote SEREs, and φ, φ_1, and φ_2 denote FL formulas:

- $\varphi_1 \vee \varphi_2 \stackrel{\text{def}}{=} \neg(\neg\varphi_1 \wedge \neg\varphi_2)$
- $\varphi_1 \to \varphi_2 \stackrel{\text{def}}{=} \neg\varphi_1 \vee \varphi_2$
- $\varphi_1 \leftrightarrow \varphi_2 \stackrel{\text{def}}{=} (\varphi_1 \to \varphi_2) \wedge (\varphi_2 \to \varphi_1)$

- $F\varphi \stackrel{\text{def}}{=} [true\ U\ \varphi]$
- $G\varphi \stackrel{\text{def}}{=} \neg F \neg\varphi$
- $X\varphi \stackrel{\text{def}}{=} \neg X!\ \neg\varphi$
- $[\varphi_1\ W\ \varphi_2] \stackrel{\text{def}}{=} [\varphi_1\ U\ \varphi_2] \vee G\varphi_1$

- $always\ \varphi \stackrel{\text{def}}{=} G\ \varphi$
- $never\ \varphi \stackrel{\text{def}}{=} G\ \neg\varphi$
- $next!\ \varphi \stackrel{\text{def}}{=} X!\ \varphi$
- $next\ \varphi \stackrel{\text{def}}{=} X\ \varphi$
- $eventually!\ \varphi \stackrel{\text{def}}{=} F\varphi$

- $\varphi_1\ until!\ \varphi_2 \stackrel{\text{def}}{=} [\varphi_1\ U\ \varphi_2]$
- $\varphi_1\ until\ \varphi_2 \stackrel{\text{def}}{=} [\varphi_1\ W\ \varphi_2]$
- $\varphi_1\ until!\_\ \varphi_2 \stackrel{\text{def}}{=} [\varphi_1\ U\ \varphi_1 \wedge \varphi_2]$
- $\varphi_1\ until\_\ \varphi_2 \stackrel{\text{def}}{=} [\varphi_1\ W\ \varphi_1 \wedge \varphi_2]$

- $\varphi_1\ before!\ \varphi_2 \stackrel{\text{def}}{=} [\neg\varphi_2\ U\ \varphi_1 \wedge \neg\varphi_2]$
- $\varphi_1\ before\ \varphi_2 \stackrel{\text{def}}{=} [\neg\varphi_2\ W\ \varphi_1 \wedge \neg\varphi_2]$
- $\varphi_1\ before!\_\ \varphi_2 \stackrel{\text{def}}{=} [\neg\varphi_2\ U\ \varphi_1]$
- $\varphi_1\ before\_\ \varphi_2 \stackrel{\text{def}}{=} [\neg\varphi_2\ W\ \varphi_1]$

- $X!\ [i]\varphi \stackrel{\text{def}}{=} \overbrace{X!\ X!\ ...X!}^{i\ times}\ \varphi$
- $X[i]\varphi \stackrel{\text{def}}{=} \overbrace{XX...X}^{i\ times}\varphi$
- $next![i]\ \varphi \stackrel{\text{def}}{=} X!\ [i]\ \varphi$
- $next[i]\ \varphi \stackrel{\text{def}}{=} X[i]\ \varphi$
- $next\_a![i..j]\varphi \stackrel{\text{def}}{=} (X![i]\varphi) \wedge \ldots \wedge (X![j]\varphi)$
- $next\_a[i..j]\varphi \stackrel{\text{def}}{=} (X[i]\varphi) \wedge \ldots \wedge (X[j]\varphi)$
- $next\_e![i..j]\varphi \stackrel{\text{def}}{=} (X![i]\varphi) \vee \ldots \vee (X![j]\varphi)$
- $next\_e[i..j]\varphi \stackrel{\text{def}}{=} (X[i]\varphi) \vee \ldots \vee (X[j]\varphi)$

- $next\_event!(b)(\varphi) \stackrel{\text{def}}{=} [\neg b\ U\ b \wedge \varphi]$
- $next\_event(b)(\varphi) \stackrel{\text{def}}{=} [\neg b\ W\ b \wedge \varphi]$
- $next\_event!(b)[k](\varphi) \stackrel{\text{def}}{=}$
  $next\_event!(b)\ \overbrace{(X!\ next\_event!(b)...(X!\ next\_event!(b)}^{k-1\ times}(\varphi))...)$
- $next\_event(b)[k](\varphi) \stackrel{\text{def}}{=}$
  $next\_event(b)\ \overbrace{(Xnext\_event(b)...(Xnext\_event(b)}^{k-1\ times}(\varphi))...)$
- $next\_event\_a!(b)[k..l](\varphi) \stackrel{\text{def}}{=} next\_event!(b)[k](\varphi) \wedge ... \wedge next\_event!(b)[l](\varphi)$
- $next\_event\_a(b)[k..l](\varphi) \stackrel{\text{def}}{=} next\_event(b)[k](\varphi) \wedge ... \wedge next\_event(b)[l](\varphi)$
- $next\_event\_e!(b)[k..l](\varphi) \stackrel{\text{def}}{=} next\_event!(b)[k](\varphi) \vee ... \vee next\_event!(b)[l](\varphi)$
- $next\_event\_e(b)[k..l](\varphi) \stackrel{\text{def}}{=} next\_event(b)[k](\varphi) \vee ... \vee next\_event(b)[l](\varphi)$

- $r(\varphi) \stackrel{\text{def}}{=} r \mapsto \varphi$
- $r \Rightarrow \varphi \stackrel{\text{def}}{=} \{r; true\} \mapsto \varphi$

- φ abort $b \stackrel{\text{def}}{=} \varphi$ async_abort b

B.4.3 Parameterized SEREs and formulas

Let r be a SERE, and l, m be integers. Let S be a set of constants, integers or boolean values and p an identifier. The left-hand side of the following are SEREs, equivalent to the SEREs on the right-hand side:

- for $p\ in\ S : |\ r \stackrel{\text{def}}{=} \underset{s \in S}{|} \{r[p \leftarrow s]\}$.
- for $p\langle l..m\rangle\ in\ S : |\ r \stackrel{\text{def}}{=} \underset{s_l \in S}{|} \ldots \underset{s_m \in S}{|} \{r[p\langle l..m\rangle \leftarrow \langle s_l..s_m\rangle]\}$
- for $p\ in\ S : \&\&\ r \stackrel{\text{def}}{=} \underset{s \in S}{\&\&} \{r[p \leftarrow s]\}$.

- for $p\langle l..m\rangle \; in \; S : \&\& \; r \stackrel{\text{def}}{=} \underset{s_l \in S}{\&\&} \cdots \underset{s_m \in S}{\&\&} \{r[p\langle l..m\rangle \leftarrow \langle s_l..s_m\rangle]\}$
- for $p \; in \; S : \& \; r \stackrel{\text{def}}{=} \underset{s \in S}{\&} \{r[p \leftarrow s]\}$.
- for $\& \; p\langle l..m\rangle \; in \; S : \& \; r \stackrel{\text{def}}{=} \underset{s_l \in S}{\&} \cdots \underset{s_m \in S}{\&} \{r[p\langle l..m\rangle \leftarrow \langle s_l..s_m\rangle]\}$

where $r[p \leftarrow s]$ is the SERE obtained from r by replacing every occurrence of p by s and $r[p\langle l..m\rangle \leftarrow \langle s_l..s_m\rangle]$ is the SERE obtained from r by replacing every occurrence of p_j with s_j for all j such that $l \leq j \leq m$.

Let f be a PSL formula, and l, m integers. Let S be a set of constants, integers or boolean values and p an identifier. The left-hand side of the following are PSL formulas equivalent to the PSL formulas on the right-hand side:

- for $p \; in \; S : \vee \; f \stackrel{\text{def}}{=} \bigvee_{s \in S} f[p \leftarrow s]$
- for $p\langle l..m\rangle \; in \; S : \vee \; f \stackrel{\text{def}}{=} \bigvee_{s_l \in S} \cdots \bigvee_{s_m \in S} f[p\langle l..m\rangle \leftarrow \langle s_l..s_m\rangle]$
- for $p \; in \; S : \wedge \; f \stackrel{\text{def}}{=} \bigwedge_{s \in S} f[p \leftarrow s]$
- for $p\langle l..m\rangle \; in \; S : \wedge \; f \stackrel{\text{def}}{=} \bigwedge_{s_l \in S} \cdots \bigwedge_{s_m \in S} f[p\langle l..m\rangle \leftarrow \langle s_l..s_m\rangle]$
- forall $p \; in \; S : \; f \stackrel{\text{def}}{=}$ for $p \; in \; S : \wedge \; f$
- forall $p\langle l..m\rangle \; in \; S : \; f \stackrel{\text{def}}{=}$ for $p\langle l..m\rangle \; in \; S : \wedge \; f$

where $f[p \leftarrow s]$ is the formula obtained from f by replacing every occurrence of p by s and $f[p\langle l..m\rangle \leftarrow \langle s_l..s_m\rangle]$ is the formula obtained from f by replacing every occurrence of p_j with s_j for all j such that $l \leq j \leq m$.

B.5 Rewriting rules for clocks

In Section B.3.1, we gave the semantics of clocked FL formulas directly. There is an equivalent definition in terms of unclocked FL formulas, as follows: Starting from the outermost clock, use the following rules to translate clocked SEREs into unclocked SEREs, and clocked FL formulas into unclocked FL formulas.

The rewrite rules for SEREs are:

1. $\mathcal{R}^c(\{r\}) = \mathcal{R}^c(r)$
2. $\mathcal{R}^c(b) = \neg c[*]; c \wedge b$
3. $\mathcal{R}^c(r_1 \; ; \; r_2) = \mathcal{R}^c(r_1) \; ; \mathcal{R}^c(r_2)$

4. $\mathcal{R}^c(r_1 : r_2) = \{\mathcal{R}^c(r_1)\} : \{\mathcal{R}^c(r_2)\}$
5. $\mathcal{R}^c(r_1 \mid r_2) = \{\mathcal{R}^c(r_1)\} \mid \{\mathcal{R}^c(r_2)\}$
6. $\mathcal{R}^c(r_1 \ \&\& \ r_2) = \{\mathcal{R}^c(r_1)\} \ \&\& \ \{\mathcal{R}^c(r_2)\}$
7. $\mathcal{R}^c([*0]) = [*0]$
8. $\mathcal{R}^c(r[*]) = \{\mathcal{R}^c(r)\}[*]$
9. $\mathcal{R}^c(r@c_1) = \mathcal{R}^{c_1}(r)$

The rewrite rules for FL formulas are:

1. $\mathcal{F}^c((\varphi)) = (\mathcal{F}^c(\varphi))$
2. $\mathcal{F}^c(b!) = [\neg c \ U \ (c \wedge b)]$
3. $\mathcal{F}^c(b) = [\neg c \ W \ (c \wedge b)]$
4. $\mathcal{F}^c(\neg\varphi) = \neg\mathcal{F}^c(\varphi)$
5. $\mathcal{F}^c(\varphi \wedge \psi) = (\mathcal{F}^c(\varphi) \wedge \mathcal{F}^c(\psi))$
6. $\mathcal{F}^c(X!\varphi) = [\neg c \ U \ (c \wedge X! \ [\neg c \ U \ (c \wedge \mathcal{F}^c(\varphi))])]$
7. $\mathcal{F}^c(\varphi \ U \ \psi) = [(c \rightarrow \mathcal{F}^c(\varphi)) \ U \ (c \wedge \mathcal{F}^c(\psi))]$
8. $\mathcal{F}^c(r \mapsto \varphi) = \mathcal{R}^c(r) \mapsto \mathcal{F}^c(\varphi)$
9. $\mathcal{F}^c(r!) = \mathcal{R}^c(r)!$
10. $\mathcal{F}^c(r) = \mathcal{R}^c(r)$
11. $\mathcal{F}^c(\varphi@c_1) = \mathcal{F}^{c_1}(\varphi)$
12. $\mathcal{F}^c(\varphi$ async_abort $b) = \mathcal{F}^c(\varphi)$ async_abort b
13. $\mathcal{F}^c(\varphi$ sync_abort $b) = \mathcal{F}^c(\varphi)$ sync_abort $(b \wedge c)$

C

Operator Precedence

The table below gives the order of precedence of the operators as well as their associativity. Here `next*` and `next_event*` stand for all the variations of the `next` and `next_event` operators, and `until*` and `before*` stand for all the variations of the `until` and `before` operators.

<table>
<tr><th></th><th>Operator</th><th>Associativity</th></tr>
<tr><td>High</td><td>HDL operators of the base flavor (e.g. && and ||) according to their precedence in the base flavor</td><td></td></tr>
<tr><td></td><td>union</td><td>left</td></tr>
<tr><td></td><td>@</td><td>left</td></tr>
<tr><td></td><td>[*] [+] [=] [->]</td><td>left</td></tr>
<tr><td></td><td>within</td><td>left</td></tr>
<tr><td></td><td>& && (SERE and's)</td><td>left</td></tr>
<tr><td></td><td>| (SERE or)</td><td>left</td></tr>
<tr><td></td><td>:</td><td>left</td></tr>
<tr><td></td><td>;</td><td>left</td></tr>
<tr><td></td><td>abort async_abort sync_abort</td><td>left</td></tr>
<tr><td></td><td>next* next_event* eventually!</td><td>right</td></tr>
<tr><td></td><td>until* before*</td><td>right</td></tr>
<tr><td></td><td>|-> |=></td><td>right</td></tr>
<tr><td></td><td>-> <-></td><td>right</td></tr>
<tr><td>Low</td><td>always never</td><td>right</td></tr>
</table>

D

Quick Reference

D.1 Logical operators

D.1.1 Verilog, SystemVerilog and SystemC flavors

Here b is a Boolean expression, p, q properties, L a list of values, j, k integers, and x an identifier; p(x) indicates a property p that uses identifier x.

Property	Intuitive Meaning
b	b holds at the current cycle
!p	p does not hold at the current cycle
p && q	both p and q hold at the current cycle
p || q	either p or q holds at the current cycle
p -> q	if p holds at the current cycle then q holds at the current cycle as well
p <-> q	shortcut for (p -> q) && (q -> p)
for x in boolean: && p(x)	shortcut for p('true) && p('false)
for x in {L}: && p(x)	shortcut for p(l$_1$) && p(l$_2$) && ... && p(l$_n$) where l$_1$, l$_2$, etc. are items from list L
for x in {j:k}: && p(x)	shortcut for p(j) && p(j+1) && ... && p(k)
for x in boolean: || p(x)	shortcut for p('true) || p('false)
for x in {L}: || p(x)	shortcut for p(l$_1$) || p(l$_2$) ... || p(l$_n$) where l$_1$, l$_2$, etc. are items from list L
for x in {j:k}: || p(x)	shortcut for p(j) || p(j+1) ... || p(k)

D.1.2 Logical operators in the VHDL flavor

Here `b` is a Boolean expression, `p`, `q` properties, `L` a list of values, `j`, `k` integers, and `x` an identifier; `p(x)` indicates a property `p` that uses identifier `x`.

Property	Intuitive Meaning
`b`	`b` holds at the current cycle
`not p`	`p` does not hold at the current cycle
`p and q`	both `p` and `q` hold at the current cycle
`p or q`	either `p` or `q` holds at the current cycle
`p -> q`	if `p` holds at the current cycle then `q` holds at the current cycle as well
`p <-> q`	shortcut for `(p -> q) and (q -> p)`
`for x in boolean: and p(x)`	shortcut for `p('true) and p('false)`
`for x in {L}: and p(x)`	shortcut for p(l_1) and p(l_2) and ... and p(l_n) where l_1, l_2, etc. are items from list `L`
`for x in {j:k}: and p(x)`	shortcut for `p(j) and p(j+1) and ... and p(k)`
`for x in boolean: or p(x)`	shortcut for `p('true) or p('false)`
`for x in {L}: or p(x)`	shortcut for p(l_1) or p(l_2) ... or p(l_n) where l_1, l_2, etc. are items from list `L`
`for x in {j:k}: or p(x)`	shortcut for `p(j) or p(j+1) ... or p(k)`

D.1.3 Logical operators in the GDL flavor

Here `b` is a Boolean expression, `p`, `q` properties, `L` a list of values, `j`, `k` integers, and `x` an identifier; `p(x)` indicates a property `p` that uses identifier `x`.

<table>
<tr><th>Property</th><th>Intuitive Meaning</th></tr>
<tr><td><code>b</code></td><td><code>b</code> holds at the current cycle</td></tr>
<tr><td><code>!p</code></td><td><code>p</code> does not hold at the current cycle</td></tr>
<tr><td><code>p & q</code></td><td>both <code>p</code> and <code>q</code> hold at the current cycle</td></tr>
<tr><td><code>p | q</code></td><td>either <code>p</code> or <code>q</code> holds at the current cycle</td></tr>
<tr><td><code>p -> q</code></td><td>if <code>p</code> holds at the current cycle
then <code>q</code> holds at the current cycle as well</td></tr>
<tr><td><code>p <-> q</code></td><td>shortcut for <code>(p -> q) & (q -> p)</code></td></tr>
<tr><td><code>for x in boolean: & p(x)</code></td><td>shortcut for <code>p('true) & p('false)</code></td></tr>
<tr><td><code>for x in {L}: & p(x)</code></td><td>shortcut for
$p(l_1)$ & $p(l_2)$ & ... & $p(l_n)$
where l_1, l_2, etc. are items from list <code>L</code></td></tr>
<tr><td><code>for x in {j:k}: & p(x)</code></td><td>shortcut for
<code>p(j) & p(j+1) & ... & p(k)</code></td></tr>
<tr><td><code>for x in boolean: | p(x)</code></td><td>shortcut for <code>p('true) | p('false)</code></td></tr>
<tr><td><code>for x in {L}: | p(x)</code></td><td>shortcut for $p(l_1)$ | $p(l_2)$ ... | $p(l_n)$
where l_1, l_2, etc. are items from list <code>L</code></td></tr>
<tr><td><code>for x in {j:k}: | p(x)</code></td><td>shortcut for <code>p(j) | p(j+1) ... | p(k)</code></td></tr>
</table>

D.2 LTL style

D.2.1 always, never and eventually!

Here p and q are properties.

Property	Intuitive Meaning
always p	p holds at the current cycle and at all future cycles
never p	p does not hold at the current cycle, nor does it hold at some future cycle
eventually! p	p holds at the current cycle or at some future cycle

D.2.2 The next* operators

Here p is a property and m and n are integers such that m≥1 and n≥m.

Property	Intuitive Meaning
next! p	p holds in the next cycle, and there must be such a cycle
next p	p holds in the next cycle, if there is such a cycle
next![n] p	p holds on the nth next cycle, and there must be such a cycle
next[n] p	p holds on the nth next cycle, if there is such a cycle
`next_e![m:n](p)`	p holds on one of the next mth through nth cycles, and there must be such a cycle
next_e[m:n](p)	p holds on one of the next mth through nth cycles, if there are at least n next cycles
`next_a![m:n](p)`	p holds on all of the next mth through nth cycles, and there must be at least n next cycles
next_a[m:n](p)	p holds on all of the next mth through nth cycles, however many exist

D.2.3 The next_event* operators

Here b is a Boolean expression, p is a property, and m and n are integers such that m≥1 and n≥m.

Property	Intuitive Meaning
next_event!(b)(p)	p holds at the next cycle b holds, and there must be such a cycle
next_event(b)(p)	p holds at the next cycle b holds, if there is such a cycle
`next_event(b)![n](p)`	p holds at the nth next cycle in which b holds, and there must be such a cycle
next_event(b)[n](p)	p holds at the nth next cycle in which b holds, if there is such a cycle
next_event_e!(b)[m:n](p)	p holds at one of the next m^{th} through n^{th} cycles in which b holds, and there must be such a cycle
next_event_e(b)[m:n](p)	p holds at one of the next m^{th} through n^{th} cycles in which b holds, if there are at least n such cycles
next_event_a!(b)[m:n](p)	p holds at all of the next m^{th} through n^{th} cycles in which b holds, and there must be at least n such cycles
next_event_a(b)[m:n](p)	p holds at all of the next m^{th} through n^{th} cycles in which b holds, however many exist

D.2.4 The until* and before* operators

Here p and q are properties.

Property	Intuitive Meaning
p until! q	p holds until the cycle where q holds, and q eventually holds
p until q	p holds until the cycle where q holds; if q never holds, p holds forever (until the end of the trace)
p until!_ q	p holds until the cycle where q holds, inclusive, and q eventually holds
p until_ q	p holds until the cycle where q holds, inclusive; if q never holds, p holds forever (until the end of the trace)
p before! q	p holds strictly before the cycle where q holds, and p eventually holds
p before q	p holds strictly before the cycle where q holds; if p never holds, then neither does q
p before!_ q	p holds before or at the same cycle where q holds, and p eventually holds
p before_ q	p holds before or at the same cycle where q holds; if p never holds, then neither does q

D.2.5 Abort operators

Here b is a Boolean expression and p is a property.

Property	Intuitive Meaning
p async_abort b	either p holds or up until b holds, p does not fail; b recognized asynchronously
p sync_abort b	either p holds or up until b holds, p does not fail; b recognized with respect to current clock context
p abort b	equivalent to p async_abort b

D.2.6 LTL operators

The Foundation Language is based on the temporal logic LTL. PSL supports the LTL operators shown in the table below. Here `p` and `q` are properties.

LTL operator	Synonym for
`G p`	`always p`
`F p`	`eventually! p`
`X! p`	`next! p`
`X p`	`next p`
`p U q`	`p until! q`
`p W q`	`p until q`

D.3 SERE style

D.3.1 Consecutive repetition operators

Here b is a Boolean expression, s is a SERE, and i,j are integers such that i $\geq$ 0 and j $\geq$ i.

SERE	Intuitive Meaning
b[*i]	i consecutive repetitions of b
b[*i:j]	between i to j consecutive repetitions of b
b[*i:inf]	at least i consecutive repetitions of b
b[*]	zero or more consecutive repetitions of b
b[+]	one or more consecutive repetitions of b
s[*i]	i consecutive repetitions of s
s[*i:j]	between i to j consecutive repetitions of s
s[*i:inf]	at least i consecutive repetitions of s
s[*]	zero or more consecutive repetitions of s
s[+]	one or more consecutive repetitions of s
‘true	skip one cycle
[*i]	skip exactly i cycles
[*i:j]	skip between i to j cycles
[*i:inf]	skip at least i cycles
[*]	skip zero or more cycles
[+]	skip one or more cycles

D.3.2 Non-consecutive and goto repetition operators.

Here b is a Boolean expression and i,j,m,n are integers such that i $\geq$ 0, j $\geq$ i, m $\geq$ 1 and n $\geq$ m.

SERE	Intuitive Meaning
b[=i]	i not necessarily consecutive repetitions of b equivalent to {!b[*]; b}[*i];!b[*]
b[=i:j]	at least i and no more than j not necessarily consecutive repetitions of b equivalent to {!b[*]; b}[*i:j];!b[*]
b[=i:inf]	at least i not necessarily consecutive repetitions of b equivalent to {!b[*]; b}[*i:inf];!b[*]
b[->m]	m not necessarily consecutive repetitions of b, and b holds at the last cycle equivalent to {!b[*]; b}[*m]
b[->m:n]	at least m and no more than n not necessarily consecutive repetitions of b, and b holds at the last cycle equivalent to {!b[*]; b}[*m:n]
b[->m:inf]	at least m not necessarily consecutive repetitions of b, and b holds at the last cycle equivalent to {!b[*]; b}[*m:inf]
b[->]	shortcut for b[->1] equivalent to {!b[*]; b}

D.3.3 Other SERE operators

Here s and t are SEREs, L is a list of values, j and k are integers, and x is an identifier; s(x) indicates a SERE s that uses the identifier x.

<table>
<tr><th>SERE</th><th>Intuitive Meaning</th></tr>
<tr><td>s ; t</td><td>match of s followed by match of t,
t starts the cycle after s ends</td></tr>
<tr><td>s : t</td><td>match of s followed by match of t,
t starts the same cycle that s ends</td></tr>
<tr><td>s | t</td><td>match of s or match of t</td></tr>
<tr><td>s && t</td><td>match of s and match of t,
lengths are the same</td></tr>
<tr><td>s & t</td><td>match of s and match of t,
lengths may be different</td></tr>
<tr><td>s within t</td><td>match of s within sequence of cycles matching t,
shortcut for {[*] ; s ; [*]} && {t}</td></tr>
<tr><td>for x in boolean: | s(x)</td><td>shortcut for s('true) | s('false)</td></tr>
<tr><td>for x in {L}: | s(x)</td><td>shortcut for
s(l_1) | s(l_2) | ... | s(l_n)
where l_1, l_2, etc. are items from list L</td></tr>
<tr><td>for x in {j:k}: | s(x)</td><td>shortcut for
s(j) | s(j+1) | ... | s(k)</td></tr>
<tr><td>for x in boolean: && s(x)</td><td>shortcut for s('true) && s('false)</td></tr>
<tr><td>for x in {L}: && s(x)</td><td>shortcut for
s(l_1) && s(l_2) && ... && s(l_n)
where l_1, l_2, etc. are items from list L</td></tr>
<tr><td>for x in {j:k}: && s(x)</td><td>shortcut for
s(j) && s(j+1) && ... && s(k)</td></tr>
<tr><td>for x in boolean: & s(x)</td><td>shortcut for s('true) & s('false)</td></tr>
<tr><td>for x in {L}: & s(x)</td><td>shortcut for
s(l_1) & s(l_2) & ... & s(l_n)
where l_1, l_2, etc. are items from list L</td></tr>
<tr><td>for x in {j:k}: & s(x)</td><td>shortcut for
s(j) & s(j+1) & ... & s(k)</td></tr>
</table>

D.3.4 Common SERE style properties

Here `s` and `t` are SEREs, and `p` is a property.

SERE	Intuitive Meaning
`never t`	there is never a match of `t`
`s |=> t!`	*if* there is a match of `s`, *then* there is a match of `t` on the suffix of the trace • `t` starts the *cycle after* match of `s` ends • *every* match of `s` must see `t` • the match of `t` must reach its end
`s |=> t`	*if* there is a match of `s`, *then* there is a match of `t` on the suffix of the trace • `t` starts the *cycle after* match of `s` ends • *every* match of `s` must see `t` • the match of `t` may "get stuck" in the middle, for instance in a starred subsequence
`s |-> t!`	*if* there is a match of `s`, *then* there is a match of `t` on the suffix of the trace • `t` starts the *same cycle* that match of `s` ends • *every* match of `s` must see `t` • the match of `t` must reach its end
`s |-> t`	*if* there is a match of `s`, *then* there is a match of `t` on the suffix of the trace • `t` starts the *same cycle* that match of `s` ends • *every* match of `s` must see `t` • the match of `t` may "get stuck" in the middle, for instance in a starred subsequence
`s |=> p`	*if* there is a match of `s`, *then* `p` holds on the suffix of the trace • suffix starts the *cycle after* match of `s` ends • *every* match of `s` must see `p`
`s |-> p`	*if* there is a match of `s`, *then* `p` holds on the suffix of the trace • suffix starts the *same cycle* that match of `s` ends • *every* match of `s` must see `p`

D.4 Clocking

D.4.1 Clocking properties

Here `p` is a property and `c` is a Boolean expression.

Clock operator	Intuitive Meaning
`p@rose(c)`	filters out all but the cycles on which `rose(c)` holds
`p@(posedge c)`	same as `p@rose(c)`
`p@fell(c)`	filters out all but the cycles on which `fell(c)` holds
`p@(negedge c)`	same as `p@fell(c)`
`p@c`	filters out all but the cycles on which `c` holds

D.4.2 Clocking SEREs

Here `s` is a SERE and `c` is a Boolean expression.

Clock operator	Intuitive Meaning
`s@rose(c)`	filters out all but the cycles on which `rose(c)` holds
`s@(posedge c)`	same as `s@rose(c)`
`s@fell(c)`	filters out all but the cycles on which `fell(c)` holds
`s@(negedge c)`	same as `s@fell(c)`
`s@c`	filters out all but the cycles on which `c` holds

D.5 Boolean, modeling and verification layers

D.5.1 Built-in functions concerning time

Here `A` is of any type, `n` is a number, `c` is a clock expression, `b` is a bit vector and `s` is a SERE.

Built-in Function	Intuitive Meaning
`prev(A)`	value of `A` at the previous cycle with respect to its clock context
`prev(A,n)`	value of `A` at the nth previous cycle with respect to its clock context
`prev(A,n,c)`	value of `A` at the nth previous cycle with respect to clock context `c`
`next(A)`	value of `A` at the next cycle regardless of its clock context
`stable(A)`	true iff value of `A` is the same as it was at previous cycle with respect to its clock context
`stable(A,c)`	true iff value of `A` is the same as it was at previous cycle with respect to clock context `c`
`rose(b)`	true iff value of `b` is `1` and was `0` at the previous cycle with respect to its clock context
`rose(b,c)`	true iff value of `b` is `1` and was `0` at the previous cycle with respect to clock context `c`
`fell(b)`	true iff value of `b` is `0` and was `1` at the previous cycle with respect to its clock context
`fell(b,c)`	true iff value of `b` is `0` and was `1` at the previous cycle with respect to clock context `c`
`ended(s)`	true iff `s` completes at the current cycle with respect to its clock context
`ended(s,c)`	true iff `s` completes at the current cycle with respect to clock context `c`

D.5.2 Other built-in functions and the union operator

Here `A` and `B` are of any type, `n` is a number, `c` is a clock expression, `V` is a bit vector, and `L` is a list of values.

Built-in Function	Intuitive Meaning
`isunknown(V)`	true iff any bit of `V` has an unknown value
`countones(V)`	number of bits in `V` that have the value `1`
`onehot(V)`	true iff `V` contains exactly one bit with the value `1`
`onehot0(V)`	true iff `V` contains at most one bit with the value `1`
`A union B`	nondeterministic choice between `A` and `B`
`nondet({L})`	nondeterministic choice of a value from list `L`
`nondet_vector(n,{L})`	an array of length `n` whose elements are chosen nondeterministically from list `L`

D.5.3 Verification directives

Here `p` is a property, `s` is a SERE, `b` and `c` are Boolean expressions, `msg` is a string, `lname` is an identifier and `D` is a directive.

Directive	Brief Description
`assert p`	verify that `p` holds
`assert p report msg`	verify that `p` holds, report `msg` if it does not
`assume p`	constrain verification so that `p` holds
`assume_guarantee p`	constrain verification so that `p` holds, verify that `p` holds in the driving block(s)
`assume_guarantee p report msg`	constrain verification so that `p` holds, verify that `p` holds in the driving block(s), report `msg` if it does not
`restrict s`	constrain verification so that entire trace matches `s`
`restrict_guarantee s`	constrain verification so that entire trace matches `s`, verify that `s` holds in the driving block(s)
`restrict_guarantee s report msg`	constrain verification so that entire trace matches `s`, verify that `s` holds in the driving block(s), report `msg` if it does not
`cover s`	check that `s` was covered by the verification suite
`cover s report msg`	check that `s` was covered by the verification suite, report `msg` if it was
`fairness b`	constrain verification so that `b` holds infinitely many times
`strong fairness b,c`	constrain verification so that either `b` holds finitely many times or `c` holds infinitely many times
`lname: D`	identify label `lname` with `D`, and verify, constrain, etc. as per `D`

D.5.4 Verification units

Here `name` is an identifier and `mod` is a module or module instance.

Construct	Brief Description
`vunit name {...}`	group directives and modeling code
`vunit name(mod) {...}`	group directives and modeling code, bind to module `mod`
`vmode name {...}`	same as `vunit name{...}`, but cannot contain assert directives
`vmode name(mode) {...}`	same as `vunit name(mod){...}`, but cannot contain assert directives
`vprop name {...}`	same as `vunit name{...}`, but may contain only assert directives
`vprop name(mode) {...}`	same as `vunit name(mod){...}`, but may contain only assert directives

D.6 Some convenient constructs

D.6.1 Comments and Macros

Here `x` is an identifier, `L` is a statically computable list and `|L|` is the size of the list `L`.

<table>
<tr><th>Macro</th><th>Brief Description</th></tr>
<tr><td>// ... <eol></td><td>trailing comment
SystemC, SystemVerilog and Verilog flavors</td></tr>
<tr><td>-- ... <eol></td><td>trailing comment
VHDL and GDL flavors</td></tr>
<tr><td>/* ... */</td><td>block comment
SystemC, SystemVerilog, Verilog and GDL flavors</td></tr>
<tr><td>‘define, ‘ifdef</td><td>compiler directives
Verilog and SystemVerilog flavors</td></tr>
<tr><td>#define, #ifdef</td><td>compiler directives
VHDL, SystemC and GDL flavors</td></tr>
<tr><td>%for x in {L} do
...
%end</td><td>replicate the text |L| times, each time replace
the occurrence of x with an item from L
all flavors</td></tr>
<tr><td>%if expr %then
...
%else
...
%end</td><td>similar to the #if construct of the cpp preprocessor
used when encapsulated by %for
all flavors</td></tr>
</table>

D.6.2 Named properties and SEREs

Here `name` is an identifier, `type_x`,`type_y` are formal parameter types, `param1,...,paramN` are formal parameters and `actual1,...,actualN` are actual parameters.

Syntax	Brief Description
`property name(type_x param1, param2; type_y param3, ..., paramN) = property_text;`	property declaration
`name(actual1, actual2, actual3, ..., actualN);`	property instantiation
`sequence name(type_x param1, param2; type_y param3, ..., paramN) = sequence_text;`	sequence declaration
`name(actual1, actual2, actual3, ..., actualN);`	sequence instantiation

Formal parameter types

The table below gives a description of the parameter types that can be used in the declaration of a property or SERE.

Syntax	Brief Description
`boolean`	a Boolean Expression
`bit`	a single bit
`bitvector`	a vector composed of bits
`numeric`	any expression interpretable as an integer in the underlying flavor
`string`	a string
`sequence`	a braced SERE, a clocked SERE, a repeated SERE or an instantiation of a named SERE
`property`	a PSL property

D.6.3 The `forall` operator

Here `x` is an identifier, `j` and `k` are integers, and `L` a list of values; `p(x)` indicates a property `p` that uses the identifier `x`.

Syntax	Brief Description
`forall x in boolean: p(x)`	shortcut for `p('true) && p('false)`
`forall x in {L}: p(x)`	shortcut for `p(l`$_1$`) && p(l`$_2$`) && ... && p(l`$_n$`)` where `l`$_1$, `l`$_2$, etc. are items from list `L`
`forall x in {j:k}: p(x)`	shortcut for `p(j) && p(j+1) && ... && p(k)`

NOTE: In replicated properties using **`forall`**, `x` can be a vector. In such a case, each element of `x` is treated independently. For example, the property

```
forall x[0:7] in boolean:
        always ((read && data[0:7]==x[0:7]) ->
                next_event(write)(data[0:7]==x[0:7]))
```

is equivalent to the "and" of 256 properties, one for each possible value of `x[0:7]`. Similarly `x` can be a vector in parametrized properties and SEREs as well.

Bibliographic Notes

Below we give a brief history of PSL. Our aim is not to give a complete chronicle of the history of temporal logic, nor a full accounting of the history of assertions in hardware design. Furthermore, we will not list each of the many people who participated in one or more of the Accellera and IEEE committees involved in the development of PSL – their names appear in the Accellera and IEEE standards. Rather, our aim is to touch on the major milestones in the development of the language, and the main personalities and ideas that have influenced PSL from its beginnings as syntactic sugaring of the temporal logic CTL, through the move to an LTL-based paradigm, and concluding with the IEEE standardization in October 2005. For background, we include a few words about the temporal logics CTL and LTL as well.

We have made every effort to refer to all the main relevant works, however we may have missed something. If so, we apologize in advance for the omission and would welcome any corrections and/or comments.

The temporal logics LTL and CTL

The linear time logic LTL was introduced as propositional temporal logic, or PTL, by Amir Pnueli in 1977 [41], and the computation tree logic CTL was first presented by Ed Clarke and Allen Emerson in 1981 [14]. For many years, a debate as to the relative merits of each was conducted in the literature. Moshe Vardi was one of the main players in that debate – see for instance [45]. One of the main arguments is that LTL is easier to use, while CTL is easier to model check.

In 1983, Pierre Wolper argued in [46] that LTL is not expressive enough: the requirement "p holds on every even cycle" is not expressible in LTL (nor is it expressible in CTL). In fact, LTL has the expressive power of star-free regular expressions – see [21].

Development of Sugar at IBM

PSL began its life as Sugar at the IBM Haifa Research Laboratory in the early 1990's. Ilan Beer, Shoham Ben-David and Avner Landver developed Sugar as a syntactic sugaring of CTL, with the intention of making the specification process easier for users of IBM's RuleBase model checker. For instance, the `next_event` operator dates to the early days of Sugar, and `next_event(b)(f)` was at that time defined as $A[\neg b \; W \; b \wedge f]$. The concept of vacuity, about which much has been written since [8, 37, 9, 16, 42, 6, 28, 29, 44, 13], dates to these early days.

Circa 1995, regular expressions were added to the logic [10] using the syntax `{r}(p)`, where `r` is a regular expression and `p` a Sugar property, in a manner reminiscent of PDL [22]. Shortly thereafter, suffix implication – in which both the left- and right-hand sides are regular expressions – was added [7], including both weak and strong regular expressions [17]. Although the motivation was usability and not expressive power, Armoni et al. [5] showed that the addition of regular expressions has the side effect of increasing the expressive power to that of omega-regular expressions. As noted in [12], their proof, for the temporal logic ForSpec, holds for PSL as well.

Originally conceived as a language for formal verification [15, 39], 1997 saw the first use of Sugar in simulation [1].

From Accellera onwards

The Accellera FVTC (Formal Verification Technical Committee) started life in 1998 as the VFV (Verilog Formal Verification) committee of OVI (Open Verilog International). When OVI and VI (VHDL International) merged into Accellera in 2000, the charter of the committee was expanded to include VHDL in addition to Verilog. Although the name includes the term "formal verification", a single specification language for both dynamic (simulation) and static (formal) verification soon became the goal of the committee. The two of us participated in the committee from close to its inception as representatives of the candidate language Sugar.

Very important roles were played by Harry Foster and Erich Marschner, chairman and co-chairman of the FVTC. Both Harry and Erich put in an enormous amount of work behind the scenes driving the standardization process – without them it would not have happened. In addition, Erich's endless patience in hearing out the more vocal members of the committee, his care to solicit the input of the more reticent members, and his documentation of everyone's opinion was greatly appreciated by all.

Leading figures from the academic roots of PSL, Ed Clarke, Allen Emerson and Moshe Vardi, took part in the process, as did over 30 industrial representatives, including both potential users of the language as well as EDA vendors. From very early on, it was decided to choose one of a number of candidate languages as the base language to be modified and enhanced according to requirements identified by the committee. In addition to Sugar, three candidate languages were donated to Accellera for consideration: CBV from Motorola [31], represented by John Havlicek and Hillel Miller, ForSpec from Intel [5], represented by Roy Armoni, and Temporal e from Verisity [40], represented by David Van Campenhout. The committee judged the candidate languages on the basis of an extensive list of 70 requirements, and on the basis of an example document containing 74 example industry properties, expressed in each of the four languages.

The exact selection process was as follows: two candidate languages out of the four were selected by vote, after which the committee identified desired changes. The donors of the two selected languages (CBV and Sugar) then modified their original proposal as per the requested changes. The final vote, taken in April of 2002, chose Sugar (with 71.4% of the votes) to be the Accellera specification language, renamed PSL.

In between the donation of Sugar in November of 2000 and its selection by the FVTC in April of 2002, a huge amount of time was invested in conducting the technical debate in the committee. The IBM team conducting the debate consisted of the two of us as well as Shoham Ben-David. As a result of the debate, and of the changes requested by the committee during the selection process, the language underwent an evolutionary process during this time.

The most visible of the changes was the move from the branching-time semantics of CTL to the linear-time semantics of LTL, as a result of the very persuasive arguments of Moshe Vardi in favor of linear-time semantics. The work of Monika Maidl [38] was instrumental in allowing the move, as it showed that the vast majority of Sugar properties used in practice could be syntactically transformed from CTL into LTL and vice versa. This meant that while the move was deeply significant from a theoretical point of view, there was little or no impact to the user from a practical point of view, for two reasons. First, because the user's view of the language did not change – the fact that `next_event` was now defined in LTL rather than CTL was transparent to the user in the vast majority of cases (which could be ascertained on the basis of a simple syntactic test). And second, because the same tools could be used to check LTL-based Sugar as CTL-based Sugar, providing they passed the same simple syntactic test. The simple subset of PSL, described in Chapter 9, has its roots in Maidl's common fragment (see also [11]).

Two other very visible additions to the language – support for multiple clocks and the `abort` operator – are the result of requests by Intel, recalling features of its ForSpec temporal logic [5]. Some other important additions dating to this period include the flavor concept, the layered definition of the language (the original definition of Sugar did not include the modeling and verification layers), and the formal definition of finite semantics, augmenting the infinite semantics previously defined. During some of this time, the IBM team was supported by Mike Gordon, whose work on incorporating the formal semantics of PSL into HOL [25, 26, 27] uncovered some subtle bugs in the formal semantics as originally written.

The first Accellera version of PSL (PSL 1.01) [2] was released in June 2003. Accellera version 1.1 [3], released in June 2004, added a SystemVerilog flavor to the original three flavors (Verilog, VHDL, and GDL). In addition, operator precedence was overhauled and labels and report clauses for directives were added. Accellera version 1.1 also corrected three anomalies present in version 1.01. While these anomalies had minimal influence on users of the language (because they involved corner cases that tools could choose to ignore with little or no impact on the user), it was important that they ultimately be solved, because adherence to the standard is determined by adherence to the formal semantics.

The first anomaly was that originally two kinds of clocks, strong and weak, were defined, but the strength of the clock had only a minimal effect. A solution that eliminated the need for two kinds of clocks was presented in [20], and incorporated into the Accellera version 1.1 formal semantics.

The second anomaly was identified in [4], which showed that the complexity of the `abort` operator as defined in Accellera version 1.01 was problematical. A solution, based on the theory of truncated paths developed in [18], incorporated the semantics suggested by [4] but used a simpler and more elegant notation. This solution was later modified to include SEREs [19], and basic results on the resulting semantics (which were incorporated into the Accellera version 1.1 formal semantics) were documented in [30].

The third anomaly concerned weak SEREs such as {`a ; b[*] ; 'false`} (where `'false` is an expression that does not hold at any cycle), that do not match any sequence of cycles. In the formal semantics of Accellera version 1.01, such a SERE, when used as a property, would not hold on any trace, whereas the intuition and intention was that {`a ; b[*] ; 'false`}, being weak, hold on a sequence of cycles in which `a` is asserted on the first cycle and `b` on all the rest. The solution was based on the framework developed in [18, 19], and was incorporated into the Accellera version 1.1 formal semantics. However, the solution creates a new anomaly, in that it treats the *logical contradiction* `'false` differently from the *structural contradiction* {`a`} `&&` {`a;a`}. A possible solution to this was proposed in [17], which also examines in depth the issue of weak vs. strong temporal operators.

The first IEEE version (IEEE Std 1850-2005) [33] was released in October 2005. In addition to a number of clarifications on various topics, the main changes for IEEE Std 1850-2005 were the addition of a fifth flavor (SystemC), replacement of *endpoints* with the built-in function `ended()`, the addition of variations on the `abort` operator, parameterized properties and SEREs, and the introduction of the keyword `hdltype` to ease interaction with the underlying HDL.

Current status

Any attempt to list tools supporting PSL would quickly become out of date. See http://www.haifa.il.ibm.com/projects/verification/sugar/tools.html for such a list.

References

1. Y. Abarbanel, I. Beer, L. Gluhovsky, S. Keidar, and Y. Wolfsthal. FoCs: Automatic generation of simulation checkers from formal specifications. In *Proc. 12th International Conference on Computer Aided Verification (CAV 2000)*, volume 1855 of *LNCS*, pages 538–542. Springer, 2000.
2. Accellera Property Specification Language Reference Manual (version 1.01). http://www.eda.org/vfv/docs/psl_lrm-1.01.pdf.
3. Accellera Property Specification Language Reference Manual (version 1.1). http://www.eda.org/vfv/docs/PSL-v1.1.pdf.
4. R. Armoni, D. Bustan, O. Kupferman, and M.Y. Vardi. Resets vs. aborts in linear temporal logic. In *Proc. 9th International Conference on Tools and Algorithms for the Construction and Analysis of Systems (TACAS 2003)*, volume 2619 of *LNCS*, pages 65–80. Springer, 2003.
5. R. Armoni, L. Fix, A. Flaisher, R. Gerth, B. Ginsburg, T. Kanza, A. Landver, S. Mador-Haim, E. Singerman, A. Tiemeyer, M.Y. Vardi, and Y. Zbar. The ForSpec temporal logic: A new temporal property-specification language. In *Proc. 8th International Conference on Tools and Algorithms for the Construction and Analysis of Systems (TACAS 2002)*, volume 2280 of *LNCS*, pages 296–311. Springer, 2002.
6. R. Armoni, L. Fix, A. Flaisher, O. Grumberg, N. Piterman, A. Tiemeyer, and M.Y. Vardi. Enhanced vacuity detection in linear temporal logic. In *Proc. 15th International Conference on Computer Aided Verification (CAV 2003)*, volume 2725 of *LNCS*, pages 368–380. Springer, 2003.
7. I. Beer, S. Ben-David, C. Eisner, D. Fisman, A. Gringauze, and Y. Rodeh. The temporal logic Sugar. In *Proc. 13th International Conference on Computer Aided Verification (CAV 2001)*, volume 2102 of *LNCS*, pages 363–367. Springer, 2001.
8. I. Beer, S. Ben-David, C. Eisner, and Y. Rodeh. Detection of vacuity in ACTL formulas. In *Proc. 9th International Conference on Computer Aided Verification (CAV 1997)*, volume 1254 of *LNCS*, pages 279–290. Springer, 1997.
9. I. Beer, S. Ben-David, C. Eisner, and Y. Rodeh. Efficient detection of vacuity in temporal model checking. *Formal Methods in System Design*, 18(2):141–163, 2001.

10. I. Beer, S. Ben-David, and A. Landver. On-the-fly model checking of RCTL formulas. In *Proc. 10th International Conference on Computer Aided Verification (CAV 1998)*, volume 1427 of *LNCS*, pages 184–194. Springer, 1998.
11. S. Ben-David, D. Fisman, and S. Ruah. The safety simple subset. In *Proc. 1st International Haifa Verification Conference*, volume 3875 of *LNCS*, pages 14–29. Springer, 2005.
12. D. Bustan, D. Fisman, and J. Havlicek. Automata construction for PSL. Technical Report MCS05-04, The Weizmann Institute of Science, May 2005.
13. D. Bustan, A. Flaisher, O. Grumberg, O. Kupferman, and M.Y. Vardi. Regular vacuity. In *Proc. 13th Advanced Research Working Conference on Correct Hardware Design and Verification Methods (CHARME 2005)*, volume 3725 of *LNCS*, pages 191–206. Springer, 2005.
14. E.M. Clarke and E.A. Emerson. Design and synthesis of synchronization skeletons using branching-time temporal logic. In *Logic of Programs Workshop*, volume 131 of *LNCS*, pages 52–71. Springer, 1981.
15. E.M. Clarke, O. Grumberg, and D.A. Peled. *Model Checking*. MIT Press, 1999.
16. Y. Dong, B. Sarna-Starosta, C.R. Ramakrishnan, and S.A. Smolka. Vacuity checking in the modal mu-calculus. In *Proc. 9th International Conference on Algebraic Methodology and Software Technology (AMAST 2002)*, pages 147–162. Springer, 2002.
17. C. Eisner, D. Fisman, and J. Havlicek. A topological characterization of weakness. In *Proc. 24th Annual ACM SIGACT-SIGOPS Symposium on Principles Of Distributed Computing (PODC 2005)*, pages 1–8. ACM, 2005.
18. C. Eisner, D. Fisman, J. Havlicek, Y. Lustig, A. McIsaac, and D. Van Campenhout. Reasoning with temporal logic on truncated paths. In *Proc. 15th International Conference on Computer Aided Verification (CAV 2003)*, volume 2725 of *LNCS*, pages 27–39. Springer, 2003.
19. C. Eisner, D. Fisman, J. Havlicek, and J. Mårtensson. The $\top, \bot$ approach for truncated semantics. Technical Report 2006.01, Accellera, May 2006.
20. C. Eisner, D. Fisman, J. Havlicek, A. McIsaac, and D. Van Campenhout. The definition of a temporal clock operator. In *Proc. 30th International Colloquium on Automata, Languages and Programming (ICALP 2003)*, volume 2719 of *LNCS*, pages 857–870. Springer, 2003.
21. E.A. Emerson. Temporal and modal logic. In *Handbook of Theoretical Computer Science, Volume B*, chapter 16, pages 995–1072. Elsevier Science Publishers and The MIT Press, 1994.
22. M.J. Fischer and R.E. Ladner. Propositional dynamic logic of regular programs. *Journal of Computer and Systems Sciences*, 18:194–211, 1979.
23. H.D. Foster, A.C. Krolnik, and D.J. Lacey. *Assertion Based Design, 2nd Edition*. Kluwer Academic Publishers, 2004.
24. GDL – General Description Language. Available at http://standards.ieee.org/downloads/1850/1850-2005/gdl.pdf.
25. M.J.C. Gordon. Using HOL to study Sugar 2.0 semantics. In *Proc. 15th International Conference on Theorem Proving in Higher Order Logics (TPHOLs 2002, NASA/CP-2002-211736)*, pages 87–100. National Aeronautics and Space Administration, 2002.
26. M.J.C. Gordon. Validating the PSL/Sugar semantics using automated reasoning. *Formal Asp. Comput.*, 15(4):406–421, 2003.

27. M.J.C. Gordon, J. Hurd, and K. Slind. Executing the formal semantics of the Accellera Property Specification Language by mechanised theorem proving. In *Proc. 12th Advanced Research Working Conference on Correct Hardware Design and Verification Methods (CHARME 2003)*, volume 2860 of *LNCS*, pages 200–215. Springer, 2003.
28. A. Gurfinkel and M. Chechik. Extending extended vacuity. In *Proc. 5th International Conference on Formal Methods in Computer-Aided Design (FMCAD 2004)*, volume 3312 of *LNCS*, pages 306–321. Springer, 2004.
29. A. Gurfinkel and M. Chechik. How vacuous is vacuous? In *Proc. 10th International Conference on Tools and Algorithms for the Construction and Analysis of Systems (TACAS 2004)*, volume 2988 of *LNCS*, pages 451–466. Springer, 2004.
30. J. Havlicek, D. Fisman, and C. Eisner. Basic results on the semantics of Accellera PSL 1.1. Technical Report 2004.02, Accellera, May 2004.
31. J. Havlicek, N. Levi, H. Miller, and K. Shultz. Extended CBV statement semantics, partial proposal presented to the Accellera Formal Verification Technical Committee, April 2002. At http://www.eda.org/vfv/hm/att-0772/01-ecbv_statement_semantics.ps.gz.
32. IEC/IEEE Standard for Verilog Register Transfer Level Synthesis. IEC/IEEE 62142 (IEEE 1364.1™).
33. IEEE Standard for Property Specification Language (PSL). IEEE Std 1850™-2005.
34. IEEE Standard for SystemVerilog. IEEE Std 1800™-2005.
35. IEEE Standard for VHDL Register Transfer Level (RTL) Synthesis. IEEE Std 1076.6™.
36. IEEE Standard SystemC Language Reference Manual. IEEE Std 1666™-2005.
37. O. Kupferman and M.Y. Vardi. Vacuity detection in temporal model checking. In *Proc. 10th Advanced Research Working Conference on Correct Hardware Design and Verification Methods (CHARME 1999)*, volume 1703 of *LNCS*, pages 82–96. Springer, 1999.
38. M. Maidl. The common fragment of CTL and LTL. In *Proc. 41th Annual Symposium on Foundations of Computer Science (FOCS 2000)*, pages 643–652. IEEE Computer Society, 2000.
39. K.L. McMillan. *Symbolic Model Checking*. Kluwer Academic Publishers, 1993.
40. M.J. Morley. Semantics of temporal e. In *Proc. Banff'99 Higher Order Workshop (Formal Methods in Computation)*, 1999. University of Glasgow, Dept. of Computing Science Technical Report.
41. A. Pnueli. The temporal logics of programs. In *Proc. of the Annual IEEE Symposium on Foundations of Computer Science (FOCS 1977)*, pages 46–57. IEEE Computer Society, 1977.
42. M. Purandare and F. Somenzi. Vacuum cleaning CTL formulae. In *Proc. 14th International Conference on Computer Aided Verification (CAV 2002)*, volume 2404 of *LNCS*, pages 485–499. Springer, 2002.
43. RuleBase User's Manual. IBM Haifa Research Laboratory.
44. M. Samer and H. Veith. Parameterized vacuity. In *Proc. 5th International Conference on Formal Methods in Computer-Aided Design (FMCAD 2004)*, volume 3312 of *LNCS*, pages 322–336. Springer, 2004.
45. M.Y. Vardi. Branching vs. linear time: Final showdown. In *Proc. 7th International Conference on Tools and Algorithms for the Construction and Analysis of Systems (TACAS 2001)*, volume 2031 of *LNCS*. Springer, 2001.

46. P. Wolper. Temporal logic can be more expressive. *Information and Control*, 56(1/2):72–99, 1983.

Index